Contents

Case Studies in Engineering Design

Case study keywords

Foreword

This textbook is most welcome within the design community as it provides students and practising engineers with a host of practical and accessible case studies for consideration. The studies all arise from the author's extensive personal experience and are therefore representative of situations with which engineers are faced in professional life. The methodology for approaching the case studies is developed at the beginning of the book prior to application in the remaining chapters. For each case study the objectives, relevant background information and the case study tasks are presented and defined, providing the starting point for attempting the task. The text covers an attractively wide range of applications and disciplines representative of the current interdisciplinary approach to design. For the student engineer this text can be used to help stimulate skills in practical approaches to design problems. For the engineering lecturer these studies can be used as the basis for formal presentation and the setting of case studies.

Peter Childs
Lecturer in Mechanical Engineering
University of Sussex

Preface

What is the core issue of engineering design teaching? The different parts of the design process are important but perhaps the most important is the multidisciplinary nature of design. Even the simplest engineering design involves several different engineering disciplines – so it is essential that a 'multidisciplinary view' is built into the design process. One way of highlighting this approach, as part of a teaching programme, is to look not only at how products are designed but also at how their components can fail. Case studies provide a useful way of looking at this, particularly those that incorporate a range of engineering disciplines.

This is intended to be a book which can help *learning*, rather than a chronology of engineering experience. The case studies have been chosen from real life engineering design projects. Their purpose is to expose students to a wide variety of design activities and situations, including those which have incomplete or imperfect information. The student is also introduced to positive aspects such as the innovation and forward-thinking that make 'good design' what it is.

The three traditional parts of the design process: conceptual design, embodiment design and detailed engineering design, appear in many of the case studies; in some they are separated for clarity, but in others they are left 'mixed up' – the way that they often occur in the real design world. All the case studies contain a certain amount of innovation – one of the themes of this book is to encourage the student to be innovative, to try new ideas, whilst not losing sight of sound and well-proven engineering practice.

The chapters are written in a way that requires the student to perform tasks related to each case study. I believe that this is the best way of learning. The method of *approaching* design problems is all-important so different methodologies are identified and explained, in Chapter 2. These principles should not be applied rigidly to all of the case studies but are sometimes best used in parts – there are opportunities to use bits of these 'methodology' principles in all but the simplest case studies covered in the book. There are a few areas of the book where it has been necessary to sacrifice some academic rigour for practical engineering considerations. This does not mean that basic theoretical assumptions have been ignored but merely that there is not always room in each study text to include the full theoretical analysis. This can be reinforced, during lectures or classwork, as necessary.

I have tried to make this book a multidisciplinary introduction to engineering design using case studies that are interesting. If you find any errors (they do creep in) or you can see possible improvements, and certainly if you feel that the case studies are *not* interesting, then say so. You can write to me c/o the publishers at the address given on page (iv) of this book.

Clifford Matthews
BSc., C.Eng, MBA

Acknowledgements

The author wishes to express his grateful thanks to the following people and companies for their assistance towards the various case studies in this book:

Mr Hans Andren, MSc. (Civil Eng.): Design Engineer, Jacobson and Widmark AB, Lidingo, Sweden; designers of the Avesta Sculpture *God on the Rainbow*.
Mr Leslie F. E. Coombs, BSc., M. Phil. (Eng.), I.Eng., AMRAes, FRAS: Consultant ergonomist and author of *The aircraft cockpit – from stick and string to fly-by-wire*.
Mr Russel D. Haworth: Managing Director, Slingsby Aviation Ltd, York, UK.
Mr Mike Brown: Raytheon Corporate Jets Ltd, Harrow, UK.
Mr Crispin Elt, BSc.: New product design manager, John Crane Ltd, Slough, UK.
Mr K. Strom: Plant Manager, PIVCO AS, Oslo, Norway, designers of the 'Citybee' electric vehicle.
Mr L. Risidor: National Power Plc, London, UK.
Mr S. Steward: Engineering Manager, Triumph Motorcycles Ltd, Hinckley, UK.
Mr John Nadeau: Vice President, Design-style Inc., CA, USA.
Mr Peter R. N. Childs, C. Eng., M.I.Mech.E., BSc., D.Phil.: School of Engineering, University of Sussex, Brighton, UK.
Dr Gaungrui Zhu, PhD., C. Eng., M.I.Mech.E.: Senior Research Engineer, John Crane Ltd, Slough, UK.
Mr John Magnusson: President, Altech Ltd, Reykjavik, Iceland.

Mention must also be made of individuals who have provided special inspiration in one form or another: Adrian Sillitoe, Marine Engineer and absolutely the world's quickest thinker; Miles Seaman's global view of complex methodologies and other things; Ian John Mark and Ian D. Brown, engineers for whom co-operation was never too much trouble; there are many others. Special thanks are once again due to Vicky Bussell for her excellent typing of the manuscript for this book.

1 What is this book about?

This book is about design. James Watt, it is claimed, improved the design of the steam engine; likewise the Sony Corporation was credited with the invention of the 'Walkman', and Beethoven his Fifth Symphony (and no doubt his others). Perhaps these are fine historical occurrences. So, do you think that you could find the designer of Concorde, or shatterproof glass? What about the light emitting diode or nylon? The answer: probably not – in fact you would be hard pushed to find 'the designer' of the pen used to write this book. Why is this? There are several reasons, but the overriding one is the thing called *complexity*.

The discipline of design, unfortunately, seems to contain rather a lot of the world's complexity. Design is multidisciplinary, nested, sometimes intangible, repetitive and iterative. It can involve poor logic, uncertainty and paradox (often at the same time). This is where *designers* come wandering on to the scene. Often trained as single subject specialists: metallurgists, dynamicists, chemists; they enter with a flourish to address *this complexity*. Some, perish the thought, may even claim to be 'in management'. But where should they start, and how do they know when they have finished? The poetry of the answer lies in what is generally known as *the design process*; this is really what this book is about.

1.1 Engineering design territory

The territory of engineering design is simple, comprising two parts. The first, *conceptual* design – the way that something 'just has to be' – has, as its basis, inductive human thinking. It is fair to say that its solutions are helped along by a certain amount of fantasy. The second half is *practical* design, which is just the opposite; its roots are in practical and possible engineering. There are often conflicts between these two parts, so much of the engineering design 'territory' of a product is about managing imbalances between the conceptual and practical approaches. These imbalances can rarely be solved instantaneously, so, for this prime reason we can come to the early conclusion that design is a *process*. This can have several interpretations – one of the best known is the concept of 'total design' developed by Pugh (1990; ref. 1). There is nothing in any of the case studies that contradicts this 'total design' viewpoint – what the case studies do show, however, is that this design process may take many different forms.

1.2 The design process

Here is one view of the design process:

> ... well, you start with a key idea, and then throw it open for criticism. Heartened by the attention to your brainchild you endure the criticism and wait for 'the designers' to propose better ideas. You wait ... and wait ... technical specialists come and go (mainly go). Design is a bit like that ....

This has a ring of truth about it: the design process does *not* always have a clear and structured logic. Without the initial idea, however, the process cannot start – look at this important statement:

INNOVATION IS THE BEDROCK OF THE DESIGN PROCESS

This means that it is the start of the design process which is the most important, despite the interesting but occasionally circular paths that will follow. Some of the most important points made within the case studies in this book are there to help you recognise this as part of your thinking.

1.3 The purpose of this book

The chapters in this book contain case studies, designed to demonstrate to you key aspects of the design process. I have intended that the chapters should provide a structured trek through the issues so this is, in a way, a book of instructions. It cannot claim to offer a short-cut to being a successful engineering designer, nor will it provide a quick method of bluffing your way through a design course or your next product design meeting. It can, however, help you learn how to structure the design process. This will help you, because *structure* can help deal with complexity.

1.4 How to use this book

Each chapter is designed to be self-contained, providing the information and explanations necessary to help you with the accompanying case study tasks. The total inventory of cases is intended to cover a broad spread of engineering course disciplines. Suitable cases can be picked out to fit in with the specific requirements of most first degree courses containing engineering and/or design modules. Each case follows a similar approach containing background information, technical outline (and essential data where necessary), a statement of the problem and a specific set of case study tasks or exercises. Several of the cases are set within a framework of 'context information' so they are representative of the way that these design problems are experienced in real life. There is nothing contrived about any of these case studies – all have been taken from the real commercial engineering world.

Figure 1.1 gives an outline of the chapter contents and a summary of the design issues that they cover. The technical disciplines are varied, to fit in with the wide technical scope of the book. The figure also gives an indication of the level of difficulty of each case – some are specifically designed to be group exercises.

The best way to treat each chapter is first to skim briefly through the text and the figures. Try to pick out the overall flavour and direction of the design issues involved, paying particular attention to the 'problem' section. Read the 'case study task' section carefully and in depth – this will help you target the specific items of information that you need from the text. Next, read the whole case text again slowly, from start to finish, to gain a more detailed 'feel'. Pay specific attention to looking for the core information that you need to make a start on the case study tasks. Try not to get too absorbed by the descriptions that provide the 'context' of the case – but be forewarned that you cannot afford to ignore these totally; they sometimes contain pieces of key information that you will need.

A word on methodology: Chapter 2 explains problem solving methodologies, dividing them into categories for easier reference. Please do not ignore these methodologies – design can be very involved (remember *complexity*) and a good sound methodology is absolutely essential to provide structure to your activities. This means that you should read and understand Chapter 2 before attempting your first case study.

References

1.1 Pugh, S. (1990) *Total design – integrated methods for successful product engineering*. Addison-Wesley, Wokingham.

Chapter	Basic design principles	Detailed engineering design	Embodiment design	Conceptual design	Innovation	Design iteration	Standardisation	Reliability	Life assessment	Design safety	Design improvements	Failures	Ergonomics	New materials	Quality: ISO 9000	Costing/value engineering	'Total' design	Management	Reporting	Level of difficulty 1 (low) to 4 (high)
3 Crane sheave – design	●		●							●										1
4 Crane sheave – failure	●	●	●					●	●	●	●	●								1
5 Crane sheave – costing											●					●			●	1
6 Casting machine	●	●					●	●											●	2
7 Rainbow sculpture		●		●	●															2
8 Inshallah condenser												●			●			●	●	2
9 Screwed fasteners		●					●													2
10 Fasteners and couplings	●	●						●		●	●	●								2+
11 Piranha				●			●								●			●	●	3
12 Mechanical seals			●		●	●		●	●		●									3
13 Aircraft flight control	●								●				●							3
14 Power boilers								●	●	●	●	●							●	3
15 'Schloss Adler' railway			●	●						●	●									3
16 Electric vehicles			●	●		●								●		●				3
17 Motorcycles														●		●	●	●	●	4
18 FGD					●	●					●	●		●		●	●	●	●	4
19 Dying Swan yacht		●			●	●						●							●	4

Figure 1.1 Chapter content – design issues

2 Methodology

Keywords

Design problems – multidisciplinary problems – nested problems and iterative problems. Linear technical and linear procedural problems. Complexity.

2.1 About design problems

Methodology, the way in which a problem is approached, is one of the fundamentals of problem solving. It is necessary, as discussed in Chapter 1, because of complexity – a good methodology will bring structure and organisation to a problem, helping to keep this complexity to a manageable level. It can also help by guiding you towards a constructive way of thinking about a problem. A complex problem is capable of being analysed from different perspectives and a methodology can be useful in helping to eliminate some of the less constructive thought patterns. This helps clear the way for better, more incisive, ways of finding a solution.

A methodology, therefore, is a structured way of doing things – finding solutions to questions and problems of engineering design. Note also that a methodology itself has a structure (don't confuse this with the structure of the problem) which helps the methodology work – it is there only for convenience in helping to bring out a solution to a problem. To summarise, a methodology:

- is structured *for convenience*;
- consists of a *set of steps*;
- helps problem solving;
- reduces *complexity*.

Effective methodologies are generic, meaning they can be used in many 'design problem' situations and are largely independent of the specific details of that problem. They are also highly diagnostic and so encourage solutions to problems to present themselves, rather than remain hidden. Let us now look in more detail at the characteristics of design problems.

Design problems

What is it that makes design problems *special*? To answer this it is best to move backwards just a couple of steps to the origin of design problems. Where does the nature of these 'design problems' come from – does it present itself as part of the way that technology and products somehow 'develop'? The case studies in this book present some so-called problems for you to solve, but there are only a handful compared to the world's range of technologies. So, fundamentally, where does the *nature* of design problems come from?

IT COMES FROM YOU

This is what makes them special. On a basic level, however, there are a few 'common denominator' observations that can be made about design problems. These appear in many of the case studies presented in this book, and so are worth considering. They are by no means perfect, or absolute, but you may find them helpful. They are summarised below.

Design problems are multidisciplinary Even the simplest design problem normally involves several engineering disciplines. The design of, for example, a thermometer has to take into consideration strength of materials, fluid mechanics, thermodynamics and chemistry. Advanced technology designs such as aircraft involve a myriad of disciplines and principles – the list is probably endless. The key point is that discipline definitions such as fluid mechanics, thermodynamics, etc., are largely artificial – there are no discrete boundaries, as such, in the real physical world. This means that design problems are *nearly always multidisciplinary* – so it is wise to expect it. Figure 2.1 shows some of the disciplines that are involved in the design of a typical engineering component – a hydraulic shock absorber.

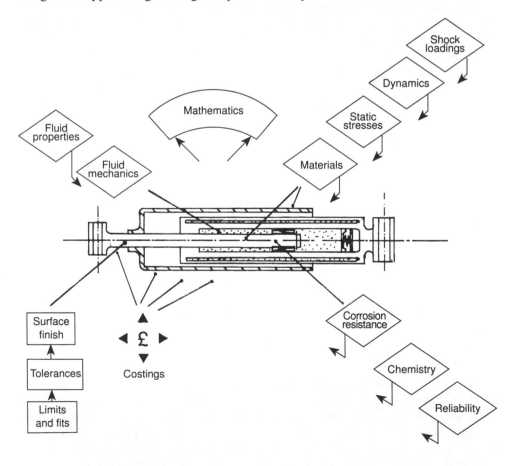

Figure 2.1 Design is multidisciplinary

Design problems are nested This is a more elusive concept to grapple with. In essence, it means that every element of a design problem contains, and is contained within, another related element of that design problem. This is another way of saying that every part of a design problem (and subsequently its solution) is affected by other parts of the problem, that themselves may not be quite so obvious. This is the property of *interrelatedness*. The concept of *nesting* is similar – it may help you to think of a design problem as a series of nested boxes or 'onion like' concentric shells. Nesting is one property that is not always easy to conceive – later we will look at specific examples to bring this concept into focus.

Design problems are iterative This in fact is not entirely accurate – it is the *solution process* which is iterative rather than the structure of the problem. The fundamental nature of design problems is such as to require an iterative process in order to generate workable solutions. Frankly, it is unusual for most real-world design problems to be solved using a purely linear methodology (i.e. a simple step-by-step way of doing things). Iteration – involving stepping back on yourself – is nearly always required, if you want to find the best solution. Once again, iteration fits in well with properties you can see at work in the real and physical world – some aspects of nature, and its evolutionary process, work like this.

Design problems then, are, by their nature, multidisciplinary, nested and iterative – a complex picture indeed. This raises the issue of one of the central messages of this book. It is:

DESIGN PROBLEMS HAVE HIGH COMPLEXITY

The concept of complexity is an important, if slightly abstract, one. If you look, you will find it in all the case studies. Complexity is to do with 'the number of states that something can be in'. The treatment of this complexity is one of the key aspects of solving design problems, of any sort.

It would not make sense to leave the subject of the nature of design problems without a glance at the availability of design *solutions*. Surprisingly, this is easier to deal with – solutions are not beset with the same problem of high complexity that we found in the design problems – solutions are simpler. There is *some* complexity but not so much that a basic toolkit of analysis and engineering skills can't deal with it. Once you have mastered these, and can understand the multidisciplinary, nested and iterative nature (those three terms again) of a design problem, then you will not have too much difficulty in finding solutions. There are a few useful guidance points to follow – these are shown in Fig. 2.2. Try to grasp the combined message behind these points rather than worrying about analysing each one individually. You will see them again, in more specific and applied form, as you work through the case studies in the book. Look at the final bullet point of Fig. 2.2 first; we will be looking at its effects later in this chapter.

Design problem types

Traditionally, books about design classify design problems by reference to their basic disciplines – hence mechanical design, dynamic design, materials of construction, and so on, are given identities and chapters of their own. Hopefully, by now, you will be able to see just how limited such an approach is. If you have had experience of

DESIGN SOLUTIONS NEED:	
• *Quality observations*	— you have to *look for* solutions.
• *Conceptualisation*	— it can help you to look for conceptual answers, before addressing all the practicalities.
• *A multidisciplinary approach*	— because multidisciplinary problems often have multidisciplinary answers.
• *Objective comparisons*	— there is rarely a single solution, so expect to have to make comparisons and *decisions*.
• *Three elements*	— good workable design solutions tend to be composed of three parts: a technical part, a procedural part and a commercial part.

Figure 2.2 Design solutions – some general points

real-world design problems you have probably found that they do not fit neatly into such simplified and discrete categories. This book takes a different approach. The case studies have been chosen to demonstrate different types of problems, granted, but the subdivisions are representative of the *fundamental nature* of the problems rather than being organised within engineering 'discipline' boundaries. The purpose of this is firstly realism, and secondly to demonstrate practical ways of categorising and then solving design problems that you will meet.

Design problems can be divided into four basic types, each with their own characteristics, both of the problem, and of the methodology that is most appropriate. It is safe to say that *all* design problems can be subdivided (quite readily with practice) into one of – or more likely some combination of – these four basic problem types. This is the first step towards neutralising some of the complexity that we know design problems possess. Let us look at these four types in turn along with their matching methodologies.

Type 1: linear technical problems
The problems This is the easiest category of problems that you will meet. They consist of a basic chain of quantitative technical steps, mainly calculations, supported by well-known engineering and physical laws. These are well accepted and therefore not in dispute. Such problems will normally have a substantial amount of 'given' information which will be readily available in a form that can be used – there will not be much uncertainty about the accuracy of this given information (it will be considered 'robust'). Figure 2.3 shows the conceptual layout of a linear technical problem. Note how the problem solving process is *linear* – each measured, quantitative substep follows on from the last on a single level – there are few, if any, iterative or retrospective viewpoints involved. Straightforward design problems of mechanics and dynamics (but not fluids or thermodynamics) often follow this kind of pattern.

Now look at the 'solution' end of Fig. 2.3. Because the solution (whatever it may be) is derived using safe, quantitative steps it will always be *checkable*, i.e. anyone can look back and check the consistency of the solution with the initial given information, and with the results of the various substeps along the way. You can think of all the steps as being technically transparent – nothing is hidden. Perhaps the

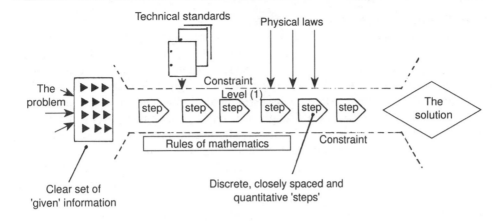

Figure 2.3 'Linear technical' problems

two key characteristics of this type of problem are this technical transparency combined with the way that the problem exists largely on a single technical level, i.e. its property of linearity.

Note that nowhere in these two characteristics is it inferred that this linear, transparent path has to be *simple* – there may be large numbers of detailed technical steps involved (a computer programming routine, for instance). The message however is that although technical content may be involved (it often is), the actual *complexity* of the problem is still low, because of the transparent quantitative steps, and their linear form.

The methodology   Linear technical problems are best handled using a methodology that mirrors the characteristics of the problem – i.e. a low complexity one. Owing to the quantitative nature of the problems, finding the technical solution becomes a mechanistic procedure. You can think of it as if the undisputed solution has in fact already been decided by the relevant theories, laws, etc., the task now being purely to find and confirm it. With this in mind, it is easier to explain the methodology as the following series of guidelines:

- You must make continuous and *rigorous* use of theory.
- Calculations (the quantitative steps) need to be accurate and consistent. 'Order of magnitude' estimates and approximations will not be good enough.
- Consult technical references – in some detail if required.
- For this type of problem, a group effort is probably *not* the best way to proceed. An individual approach is more effective, dealing with the multidisciplinary aspect as best you can.
- A linear methodology means *keep looking forward*. Start at the beginning of the problem and work through the steps to the end, checking as you go.
- Don't *invent* complexity that isn't there. Perhaps the worst effects are created by imagining all manner of procedural constraints (timescales, design margins, conventions, division of responsibilities and the like) that simply do not exist. With linear technical problems, feel justified in having some *confidence* in what you are doing.

Type 2: linear procedural problems

The problems   Just because a particular problem may be defined as *procedural* doesn't mean that it has nothing to do with engineering design. Procedural elements have an important part to play in many design situations: design costing, project management, value engineering and quantitative design techniques such as failure mode analysis (FMA) and reliability assessment are influenced by procedure. By far the best example however is the discipline of quality assurance (QA) – it relies on almost pure procedure in order to influence the design process. You will meet the techniques of QA in several of the case studies in this book.

So what does a linear procedural problem look like? Its main feature is the existence of procedural constraints, controlling what can be done both to further define the problem and, during the activities, to solve it. Do not misinterpret these as administrative constraints – they are established procedural constraints of the *technical* world. Figure 2.4 is a conceptual representation of this arrangement – note the way that the procedural constraints act to 'squeeze' the problem methodology into a linear (single-level) form. The concept of procedural constraints can be a difficult one. Consider this example: suppose Fig. 2.4 represents a problem of poor interchangeability of, for instance, motor vehicle components. The left-hand end represents the 'given' information (components for which interchangeability has been found to be poor) and the left-to-right path the solution to this problem. If you were to become involved in this problem you would find yourself beset by constraints on what you could do whilst moving (yourself) along this path. For example:

- *Measurement.* Interchangeability problems can be solved, easily, by measuring the components – but on a real vehicle assembly line there is insufficient time for 100 per cent measurement.
- *The product range.* Technically, for you, the best solution would be to make and sell only a single vehicle model, not a range with optional extras and variations. This would lessen the extent of the problem at least – but would it be commercially viable? Probably not.
- *Quality assurance.* QA principles can be used to address problems such as poor interchangeability. QA procedures and techniques however are not provided specifically for your use; they are encapsulated in systems such as the ISO 9000 series of standards which cover many different technologies – your vehicle assembly methods may prove unsuitable for implementing such a system.

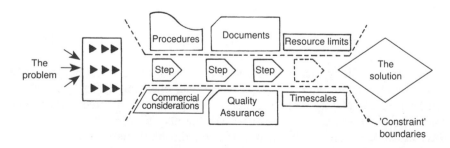

Figure 2.4 'Linear procedural' problems

Look again at these three examples of constraints – can you see how they are procedural rather than technical? They are typical of linear procedural problems.

There are analogies between this type of problem and the linear technical type discussed previously. They are still mechanistic problems; nesting is low (there are few 'problems within problems') and there can be many steps to them. There are also similarities in other areas, such as the low incidence of 'people problems' – the constraints are mainly related to documents, procedures and practices rather than the interference of other people involved in the design process.

The methodology   Once again, these are essentially problems of low complexity – even though they may look complicated. You can assume (as before) that the procedural solution has been decided for you – all you have to do is reach it. Try to follow these guidelines:

- First, check again that it really is a linear procedural problem – it must have low *people-involvement* to qualify.
- Work through the procedural steps one-by-one accepting, and responding to, the constraints that exist. You have to guard against being retrospective or you will lose confidence. Keep looking forward (left to right in Fig. 2.4).
- Use established procedural techniques such as costing methods, QA and various quantitative techniques. Don't invent your own techniques, no matter how good you think they (and you) are.
- Keep group involvement to a minimum. It really doesn't help in linear problems – there will be ample opportunity to exercise your democratic leanings on other types of problems.
- Again: don't invent complexity that doesn't exist.

Type 3: closed problems   You can be excused for feeling a little relaxed about the previous linear types of problems. This is fine; it is difficult (and wrong) to worry about problems, however well disguised, that exhibit low complexity. Unfortunately, not all problems are like this; it is necessary to learn to deal with the more complex types. The best place to start is with *closed* problems.

The problems   Closed problems actually comprise a range of problems rather than a single identifiable type. They are particularly applicable to two groups of design problems that you will meet frequently: failure investigations and new product development. Here are two closed problems:

- 'We need to design a more user-friendly keyboard'.
- 'These high-intensity halogen bulbs keep failing; we need a better design'.

The most obvious characteristic of these typical closed problems is their brevity. Look how short they are – there is no hint of the content of the 'expected' solutions or even of their form. In consequence, the technical and procedural paths to follow are something of a mystery – how do you know where to start, and when to finish? Where is the detail? Welcome to closed problems. Firstly, a point of fundamental understanding; you have seen some of this before, earlier in this chapter:

Where does the detail of closed design problems come from?

IT COMES FROM YOU

This means that a significant part of your input into such problems must be dedicated to *opening up the problem*. This is the essential step; despite the simplistic 'look' of closed problems, they only appear simple because the complexity is hidden, not because it does not exist. Closed problems, almost by definition, have *high complexity*.

Methodology   Opening up the problem is essentially an investigation process, the objective being to amplify the content and character of the problem until its complexity is revealed. Figure 2.5 shows, conceptually, how a closed problem is unfolded. This figure holds some key principles; look how the 'nesting' levels of the problem have been unfolded so that the sub-problems and sub-solutions at each level can be identified (and then dealt with).

There are a few activities which you can use to help the opening-up process:

- Look for the *common* nesting levels (see type 4 problems later). You can often anticipate these after a little practice.
- Make lists of the variables and technical parameters that you feel may be involved in the problem. This will need an appreciation of common engineering and design principles – then you will have to think for yourself in a pro-active way.
- Think *around* the problem – this means an active attempt to look for complexity (remember that you will be revealing it, not introducing it – it is there already). You can list ideas in a free-flowing brainstorming session.
- Group input will be helpful – closed problems do not respond too well to the individual approach. You will not find good solutions, or even sub-solutions, if the problem is not opened up fully. A group of minds can obtain a much *richer picture* of a problem than can one.

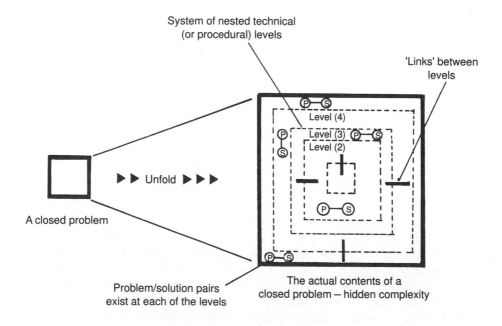

Figure 2.5 'Closed' problems

- Use estimates and 'order of magnitude' assessments rather than exact calculations.

The point has been made already that closed problems are found in a range of design situations. Many of their characteristics, however, are common, the easiest identifiable one being the characteristic of problems consisting of different 'nested' levels. Once a closed problem has been successfully opened up, the first step is to look for these component levels. The problem will then change its identity, from a closed problem to one which has been converted (consciously) into a problem with a high degree of nesting. These are known, unsurprisingly, as nested problems.

Type 4: nested problems   Unless you are blessed with an extraordinary level of intuition, or have a lot of experience, then your best chance of understanding nested design problems lies with a conceptual approach – the technique of *systems thinking*. 'Systems thinking' involves developing the ability to think of the content and format of a problem (and solution) as a series of 'boxes'. The boxes are joined to each other via various links which serve to define the interrelationships of the boxes to each other. Sometimes these links are technical, sometimes procedural. Taken together, the assembly of links and boxes comprise the problem *system* – which means that it is a conceptual representation or *model* of the problem.

The problems   There are three basic models of nested problems – together they can be joined in various ways to provide a composite model of even the most complex nested problem. Figure 2.6 shows the three models – note the different format and content of the nesting levels. Model (a) shows a problem characterised by nested technical levels. This is a common model for 'closed' failure investigation problems – you can see it in use in Chapter 19 of this book dealing with mechanical failures and redesign of a gas-turbine powered fast yacht (take a quick look forward to Fig. 19.10). The central level is referred to as 'level 1' and is completely surrounded by a second 'level 2' envelope. You can think of level 2 as containing technical issues that affect the level 1 happenings but will respond only to investigation, i.e. they will not reveal themselves unless you make a conscious effort to expose them.

Note the links that connect levels 1 and 2. These are technical links, direct connections based on proven engineering calculations, principles and laws. Level 3 is slightly different; it is at this level that the procedural aspects lie. The links between levels 2 and 3 are more difficult to define because they are neither purely technical nor procedural, but a combination of the two. The lower part of level 3 contains outside influences on the design, or the failure, as may be. Note again the main characteristic of this model; it is the nesting of *technical levels* that is the most important.

Model (b), by contrast, is an example of nested procedural levels. This is typical of problems in design and project management which contain a high degree of procedural complexity (look at the *Piranha* space vehicle project in Chapter 11). The simplest procedural aspects, such as project structure and basic design responsibilities, lie at level 1 of the model. Other relevant aspects are represented by the outer levels 2, 3 and 4 (each with progressively increasing complexity). Note the significant 'width' of these levels. It is not until level 5 that we encounter the technical issues, in this example the definition of the multiple technical interfaces

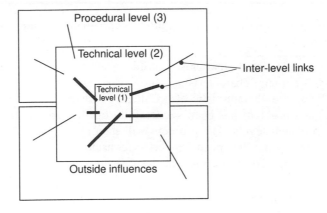

Model (a): nested technical levels

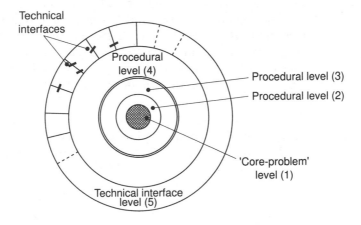

Model (b): nested procedural levels

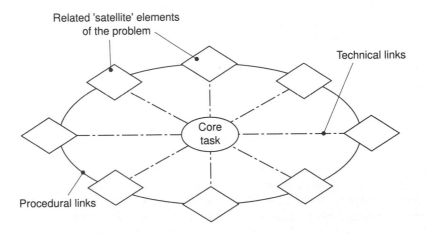

Model (c): a 'systemic problem' format

Figure 2.6 'Nested' problems

that form a part of all large and multidisciplinary design projects. Model (c) is not so straightforward; it contains a few principles which are more ingrained with the technique of systems thinking. As before, the core task, either technical or procedural, lies at the centre (level 1). This is surrounded by only one level of other problem 'satellite' elements, each relating to level 1 via a single link. Here is the key characteristic: whereas the radial links are typically technical, the circumferential links that join the level 2 satellites are *procedural*. Together these circumferential links comprise a complete and self-consistent system. This means that they are *systemic* – meaning they 'fit in' with the way that the industry, or business, or project, actually works. Do not confuse this with the term 'systematic' – which means something quite different. Several areas of design have a tendency to follow this type of conceptual model, typically:

- *creative* design activities such as graphic design and aesthetic design (of any engineering product);
- architectural design;
- computer systems analysis;
- Chapter 13, 'Aircraft flight control' in this book.

These then are the four fundamental types of problem formats, with some guidance about their particular characteristics, and matching methodologies. So what? We have seen that they are multidisciplinary, nested and iterative – and that it is possible to encapsulate them within the general concept of *complexity*. Then we have seen how it is possible to use conceptual models to enable you to think incisively about the characteristics and format of design problems. We also said (perhaps a little controversially) that the detail of a design problem is unlikely to be that clearly defined – and comes, therefore, from yourself. Does all this make design problems easier to solve? It depends on whether you think the world of engineering design is based predominantly on intuition and flair – to be more specific, that its *only* virtue is in this intuition and flair. If you do then this book (and most further education courses) are not for you. If, however, you can see the way that a structured and *organised* approach can help, you may find that the concepts put forward in this chapter will be useful to you, when you attempt the case studies in the book and later when you have to address other 'real world' design problems. Draw your own conclusions.

2.2 Key point summary

1. Methodology
- A methodology is a structured way of doing things.
- Choosing a methodology is one of the fundamentals of problem solving.

2. Design problems
- Design problems are multidisciplinary, nested (comprised of several levels) and interactive.
- Many design problems have *high complexity*.

3. Problem types

There are four basic types of design problems:

- Linear technical problems: these are the easiest, consisting of a basic chain of technical steps.
- Linear procedural problems: procedural issues are an important aspect of engineering design. This type of problem often has relatively low complexity.
- Closed problems: these are short, deceptive, and full of hidden complexity; typical examples are failure investigations and new product design.
- Nested problems: these have high complexity but can be classified into three broad types, a, b or c.

4. Design solutions

- Conceptual 'systems thinking' can help you find solutions by improving your understanding of the problem.
- Good design solutions can be quite difficult to find – you have to learn to think for yourself.

2.3 Later, back in the common room

'There's something about this that worries top students like me. All this about defining a design problem "yourself"; don't they have any objectivity?'

'Maybe, in the basic sense, however . . .'

'Oh, I see, the boundaries of a design problem can be set for you − by an examination question, or even some lecturer or design manager, I suppose − but the detail comes from you (well, probably me actually).'

'Complexity.'

'What?'

'Complexity − that's when complexity comes along and . . .'

'And you bring the complexity yourself, as part of the way you perceive things.'

'You know, I was thinking along the same lines.'

'Hey listen, I've been thinking about design, course marks and suchlike; what do you think about this? Assume that you see everything really simply; this means low complexity problems, right?'

'Right.'

'So, low complexity problems mean low complexity methodologies and solutions, easy answers, eighty five per cent marks, thank you and goodnight. What's wrong with that?'

'I'm just worried that . . . you can't make complexity . . . well, go away, just by ignoring it. Maybe it's already inherent in the design world and if you don't deal with it, then it will play with you. Anyway, if everyone ignored it then there'd only be a single style of clothes to buy, and if you wanted to buy a car, there would be no choice of model, only a single product called "car".'

'Hmm . . . no variety; that's not very *systemic*. Oh well, if we have to live with complexity then I'm sure I can think of a way to deal with it; you don't have any ideas, I suppose?'

'Actually, I do. Complexity. Complexity can deal with complexity, in fact *only* complexity can deal with complexity.'

'Profound, coming from you; I suppose you found it expressed as "so and so's law" or something.'

'Not quite, it's more of a maxim, an aphorism, that's A-P-H-O-R-I-S-M, a general truth, if you like.'

'Impressive; what about those conceptual models of problems; you're not telling me that they can address all the complexity of real design problems − just think, there must be thousands of different types, and only four models. They can't possibly represent them all.'

'Yes and no − oh, but didn't he mention the concept of usefulness?'

'Usefulness, thanks − but look, if I find model type 3 useful, chances are you probably won't; you'll prefer type 2 or type 1; that's the way the world works.'

'A little simplistic.'

'Who are you calling simple?'

'Hold on, I was referring to *using* the problem models; maybe you can fit one of the models to your real-life design problem.'

'Sorry to disagree but that's obviously not it, it's the other way round; you have to fit the real-life *problem* into the most suitable *model*.'

'And that can help you understand it?'

'Give me strength.'

'I just have.'

3 Crane sheave – basic design principles

Keywords

Good design principles – economic design – avoiding failures. Principles: short force paths – uniform strength – balanced forces – anticipating deformations. Design optimisation and choice.

3.1 Objectives

This is an exercise in not developing bad habits. It is about good design principles. Design is multidisciplinary, involving theories and rules from the worlds of thermodynamics, fluids, statics, dynamics and others. Some basic principles however are very common, appearing in most mechanical designs, whatever their particular discipline 'mix' or specialism. The good thing is that these principles are simple – they are also worth learning, not because they will guarantee you a good design but because of the way in which they can help you avoid producing very bad ones.

3.2 The crane sheave hanger

Cranes use a system of pulley mechanisms to help lift the load. The pulleys may be single or ganged depending on the application. Pulleys are located within a fabricated steel sheave – an arrangement which holds the pulley spindles (or 'pins') so that the pulleys can revolve. The simple design shown in Fig. 3.1 uses a 'hanger-type' sheave – this is common on small utility cranes up to a safe working load (SWL) of 5 tonnes. The hanger is bent from low carbon steel sheet and fits over a machined pin which is supported by two frame plates.

This design looks straightforward enough. We will now look at five basic design principles:

- economic functions;
- force transmission;
- force balance;
- uniform strength;
- matched deformations.

Try to follow the descriptions of these five design principles. It is the idea that it is important – they are outlined in general engineering terms, for similar applications, but not in the context of the crane sheave hanger (that is your job).

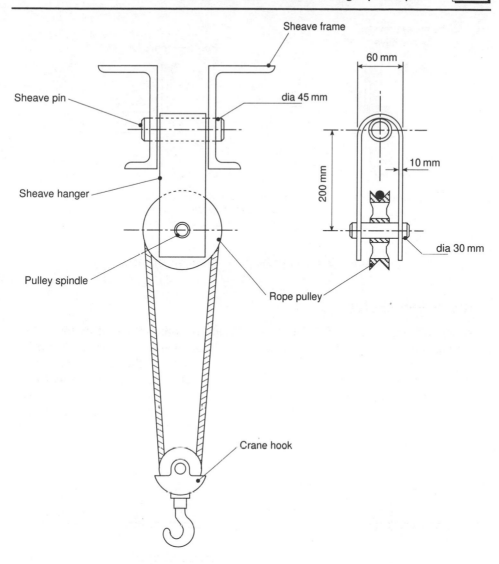

Figure 3.1 The crane sheave assembly

Economic functions

This is a poor but commonly used title. It means: pay attention to the division of tasks within the design. The objective is to make sure that each component has a well-defined task and each task is allocated to the component that can best do that task (not one of the others). The next step is to look for areas of task economy, where a single component part can do more than one task. Figure 3.2 shows how this division of tasks works in practice, for a V-belt driven shaft. The big danger in deciding division of tasks is over-complication; it is always necessary to optimise and sometimes compromise design features to stop this happening.

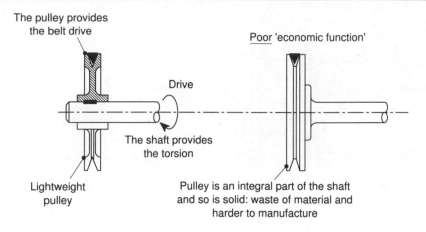

Figure 3.2 Basic principle – division of tasks

Force transmission

Most engineering designs involve the transmission of force in one of its forms. The definitions are not too rigid; force caused by a simple loading can manifest itself as complex torsional, shear or bending forces as well as the simple tensile or compressive cases. The principle is that force is best transmitted in *straight lines* (you can think of these as force 'flowlines' through a component). This is achieved by avoiding offset loadings, large deflections and sharp changes of cross-section, all of which can concentrate these 'lines of force', causing stress concentrations. Figure 3.3 shows the principle for a simple machine mounting.

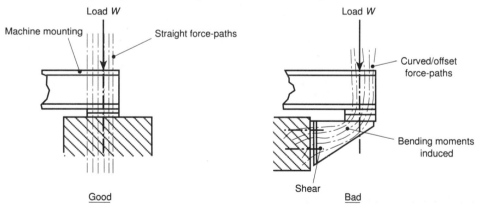

Figure 3.3 Basic principle – force transmission

Force balance

This is relevant to both static and dynamic components and mechanisms. The principle is that forces should be *balanced* so that resultant forces such as end-thrusts, imbalance and strong bending moments are, as far as is possible, eliminated.

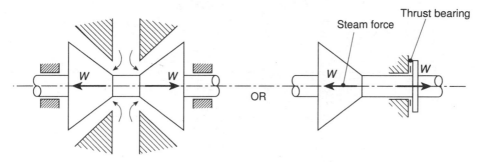

Steam turbine arrangements

Force balanced by symmetrical design   Force balanced by a thrust bearing

Figure 3.4 Basic principle – force balance

Forces are best balanced at origin, for example by aiming for design symmetry wherever possible, particularly for rotating parts. The second alternative is to ensure that residual forces are balanced by features such as collars, stiffeners or adequate section thickness as near their plane of origin as possible. Figure 3.4 shows two examples of force balancing.

Uniform strength

The principle of uniform strength is the core of the more general idea of design economy – the most economical use of materials (and therefore physical size and weight) whilst still fulfilling the duty of the design. A design which has true uniform strength has all the components stressed to the same extent and they all have the same design factor of safety. Practically, this often means that mechanisms need to be made of a selection of shapes and materials. Although a valid objective, the principle of uniform strength is often difficult to achieve in practice – sometimes it is only possible to limit the range of stresses, rather than equalise them, because of other considerations (see Fig. 3.5).

Matched deformations

This one is more difficult to understand. The objective is first to anticipate the distortions and deflections (both classed as 'deformations') that happen to an engineering component in use. The principle of good design is to match these deformations in adjacent and locating components so that they have, as far as possible, the same magnitude and sense. You can see this in operation in Fig. 3.6. The purpose of the hub/shaft arrangement is to transmit torsion, in this case from the hub, which is being driven by the other hub half, to the shaft. In use, both hub and shaft will deform, mainly in torsion. On start-up the hub will experience an initial torsional deformation which will then decrease as the torque is transferred to the shaft. To follow the principle of matched deformations the torsional stiffness of the hub needs to be matched to that of the shaft. This will make sure that the torque is

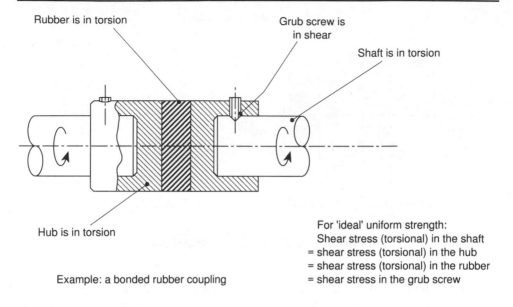

Rubber is in torsion

Grub screw is in shear

Shaft is in torsion

Hub is in torsion

For 'ideal' uniform strength:
Shear stress (torsional) in the shaft
= shear stress (torsional) in the hub
= shear stress (torsional) in the rubber
= shear stress in the grub screw

Example: a bonded rubber coupling

Figure 3.5 Basic principle – uniform strength

transferred evenly along the whole length of the shrink fit, so avoiding undesirable stress concentrations. Figure 3.6 shows good and bad design practice for the hub/shaft assembly – similar principles can be applied to other drive mechanism components such as bearings, shafts, clutches and brakes.

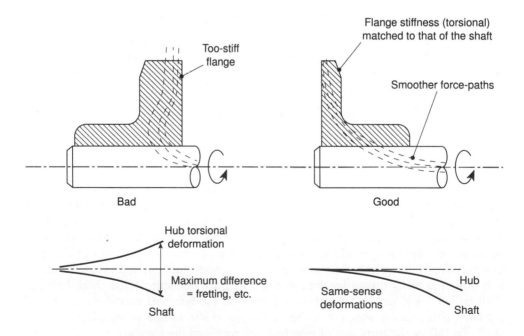

Too-stiff flange

Flange stiffness (torsional) matched to that of the shaft

Smoother force-paths

Bad

Good

Hub torsional deformation

Maximum difference = fretting, etc.

Shaft

Same-sense deformations

Hub

Shaft

Figure 3.6 Basic principle – matched deformations

Optimisation

You will often hear optimisation quoted as one of the fundamental objectives of good design. It generally involves minima; trying to obtain minimum weights, number of parts, size, and of course cost. This is fine. It is good design to try to optimise the five basic principles outlined. Conscious attempts to optimise, for instance, force balance and uniform strength will generally result in better, more balanced designs. Sometimes, however, it is not so straightforward; the principle of matched deformations does not always sit easily with the idea of keeping costs down. Compromise is often necessary – but this is not an excuse for lax treatment of these design principles.

3.3 Case study task

Figure 3.1 shows the general arrangement of the simple crane sheave described at the beginning of the case study. It is an established design which is in common use. Your task is to look at the design of the sheave hanger assembly and identify how the five simple design principles described in the study text have been used. See if you can find, in addition, evidence of any obvious compromises that have been made, perhaps because of cost, or the practicalities of manufacturing the parts. Show the application of the principles by annotating a sketch, or sketches, of the assembly and making brief notes. Detailed discussions or explanations are not necessary.

4 Crane sheave – early failures

Keywords

The failure 'event' – early conclusions – the investigation procedure – operations and design review – drawing conclusions.

4.1 Objectives

Unfortunately, even the best-designed engineering products can suffer failures early in their lives either as a result of inadequacies in the design itself or caused by external conditions and events. Failures are not easy to analyse and the precise mechanism of failure can often be difficult to identify. The objective of this case study is to look briefly at what happens when a component fails and to introduce the activity of *redesign* necessary to stop the failure recurring. It uses the same component as the previous chapter, the crane sheave, and shows how some simple design improvements are used, following a design review and analysis of the failure. This is a straightforward example – but the general principles are relevant to all mechanical component failures. More complex failure cases are included in later chapters of this book.

4.2 The failure

The crane operator instantly got most of the 'shop floor' blame. It was decided that he must have tried to lift a load which was too heavy – or if he hadn't done that, then he must have jerked the load off the shop floor. Everyone knew someone who had definitely seen the load swing wildly, and a few people who swore that it hadn't. Versions of the failure event itself varied from the sparse to the colourful; from a minor failure to total disintegration – and all were united in agreement that it was somehow the fault of 'higher management' and that it was *they* who should be thankful that no-one had been injured.

The 5 tonne overhead crane had failed, dropping its 4 tonne load of steel girders from a height of 2 metres on to the factory floor. It was clear that the failure was limited to the sheave hanger, the U-shaped bent steel bracket connecting the main rope pulley to the sheave pin and frame (Fig. 4.1). Examination of the broken parts showed that the sheave hanger had fractured horizontally along the centreline of its bend, breaking cleanly into two parts. The sheave frame, pin and pulleys were undamaged. The fracture had happened without any warning – the load had not slipped, or jerked, and all the indications were that the crane was being operated correctly, within its safe working load capability, at the time of the failure. The crane

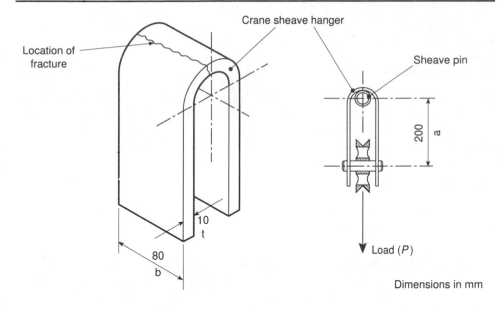

Figure 4.1 The crane sheave hanger

was nineteen months old and had recently been inspected by an independent inspection authority (a statutory requirement) and its safety certificate renewed. The design of the crane is well proven, complying with British Standard BS 466 (ref.1) and rated for a design lifetime of 2×10^6 cycles.

4.3 The investigation

A small team was assembled to investigate the failure. It was agreed that the objective was to find what had caused the failure and to take any necessary actions to prevent a similar thing happening again. The investigators fell quickly into three 'camps'; the more learned members felt that there had to be something wrong with the crane design, whilst the design-oriented people in the team felt that it was more likely to be caused by incorrect operation. The third group seemed willing to agree to almost any conclusion, as long as the failure wasn't thought to be their fault. The first suggestion of the team was to invite a representative of the crane manufacturer to come along to the initial meeting. Several telephone calls to the company only confirmed the fact that no one was available for a meeting on *that* day; however they could send a spares price list so 'you can buy some spare parts to replace those that you have broken'.

Setting the scene

The first investigation meeting convened (without the crane manufacturer or an agenda) a week after the failure. The Works Manager took charge and after giving an initial summary of the failure produced a plan of attack containing five steps:

- operations review

- design check
- examination of failed parts
- conclusions
- recommendations.

The team seemed to agree — at least no-one put forward any better ideas. Responsibilities were decided for working on the first three elements: operations review, design check and examination of failed parts. Everybody agreed that reports should be short, to the point, and not jump to instant conclusions.

The operations review

There are three things to consider during an operations review: working load, dynamic forces and fatigue life.

Working load   The crane was found to be correctly rated for a SWL of 5 tonnes, meaning that it was tested to 125 per cent SWL during its pre-commissioning proof load test. Investigations showed that the weight being lifted when the failure happened was 4 tonnes, and that no heavier weights existed on the factory floor, so the crane could not have been overloaded.

Dynamic forces   An analysis was made to see if dynamic forces could have caused overstressing of the crane components. Two movements were considered: swinging and jerking. Figure 4.2 shows the situation; swinging of the load in either the x or y plane does not exert significantly more force as the assembly is free to move in either of these planes, a stress concentration of perhaps 1.2 or 1.3 being the maximum that could be expected. Jerking the load off the ground at the full hoisting speed would cause a significant dynamic effect — it is estimated that this could be up to a factor of 4. Combining these two, as a 'worst case' combination of events, leads to a minimum required factor of safety of about 5, to cater for dynamic forces.

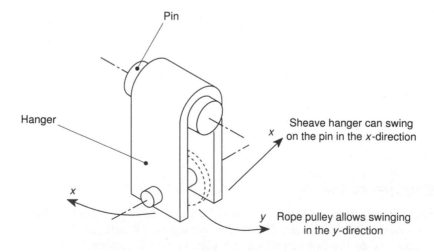

Figure 4.2 The effects of a swinging load

Fatigue life   The crane is based on a 2×10^6 cycles fatigue life, one of the design life categories mentioned in BS 466. Calculation of expired fatigue life is as follows:

$$\text{Design lifetime} = 2 \times 10^6 \text{ cycles over 25 years} = 220 \text{ cycles/day}$$
$$= 110 \text{ lifts/day}$$
$$\text{Expired lifetime} = (19 \times 30) = 570 \text{ days} \times 220 = 125\,400 \text{ cycles}$$

$$\text{Estimated life expiry} = \frac{125\,400}{2 \times 10^6} = 6.3\%$$

From this approximate calculation it seems unlikely that the crane component failed directly as a result of the type of fatigue loading for which it was designed. This assumes of course that actual stresses experienced by the material are not higher than the 'fatigue limit' used in its design calculations.

The design check

A design check concentrates on the stresses experienced in the failed component – the crane sheave hanger. The best way to approach this is to analyse the situation for an ideal stress case and then take a more pessimistic view, anticipating the effects of any *worst case* practical loading condition. Figure 4.3 shows the two anticipated states. Figure 4.3(a) is the idealised loading condition in which the sheave hanger's internal radius fits snugly over the surface of the sheave pin. This produces pure tensile stress on the hanger legs as follows:

$$\sigma \text{ tensile} = \frac{P/2}{bt} = \frac{4000 \times 9.81}{2 \times 0.08 \times 0.01} = 24.52 \text{ MN/m}^2$$

For a medium carbon steel with a yield stress R_e of about 300 MN/m² this gives a nominal factor of safety of 12. This is higher than that needed to allow for dynamic loadings – so in theory, the failure should not have occurred.

Figure 4.3(b) looks at an actual loading condition where the sheave hanger radius does not accurately fit the sheave pin; the main contact is over the top 20–30°, leaving a clearance in the horizontal plane. This changes the loading condition significantly. The load (P) now causes a bending moment. You can calculate this from standard theory:

$$\text{Maximum bending moment } M = \frac{PR^2(2-\pi)}{a+2\pi R} + \frac{PR}{2} = PR\left(\frac{R(2-\pi)+\frac{1}{2}}{a+2\pi R}\right) \text{ Nm}$$

Hence

$$M = 4000 \times 9.81 \times 0.0275 \left[\frac{0.0275(2-\pi)}{0.2+(2\pi \times 0.0275)} + \frac{1}{2}\right]$$

$$= 1079.1(-0.0842 + \tfrac{1}{2})$$

Solving:

$$M = 448.7 \text{ Nm}$$

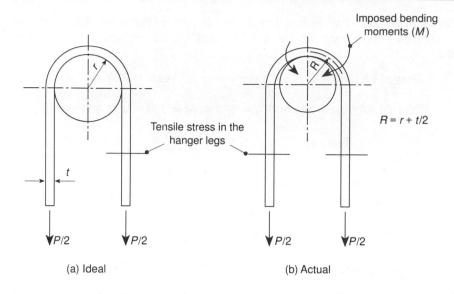

Figure 4.3 Ideal and actual loading conditions

Solving for maximum bending stress using $\sigma_b = M/z$ where $z = bt^2/6$:

$$\sigma_b = \frac{448.7}{0.08 \times 0.01^2/6} = 336.5 \text{ MN/m}^2$$

Note that this is higher than the nominal yield stress (R_e). So, even taking into account the uncertainties and assumptions of the above calculation, it is likely that the existence of such a bending moment will provide the conditions for failure, particularly when combined with the dynamic loads and fatigue effects.

Examination of failed parts

The physical examination of failed metal components follows a well-established routine, comprising both quantitative and qualitative assessment techniques. There are three main steps (refer to fig. 4.4) which are now described.

Compliance with material specification The most obvious check is to make sure that the failed component was made from the correct specified material. Elemental chemical analysis is checked using a spectrometer and mechanical properties of tensile strength, ductility and impact resistance are verified by doing a series of destructive tests on separate test pieces machined from the failed component. The results for the crane sheave hanger were quite clear – the material *was* the correct grade of low/medium carbon steel, heat-treated as specified.

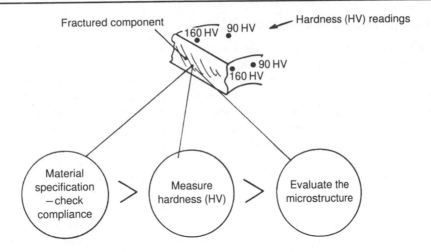

Figure 4.4 Examination of failed parts – the three steps

Material hardness Hardness is measured using the Vickers (HV) or Brinell (HB) hardness scale – a small diamond or ball being used to indent the material. The measurement gives some indication of the material's *microstructure*. The objective is to look for excessive hardness; hardness normally means brittleness, and brittle materials are prone to failure. Figure 4.4 shows the HV results in two directions relative to the fracture plane of the sheave hanger. The main observation is:

- The material is *harder* in the region of the failure (160 HV compared with 90 HV in the 'unaffected' area).

In itself, this finding is not particularly conclusive. You would expect that the plastic distortion that precedes failure of a material would cause work-hardening of the steel structure and the increase from 90 HV to 160 HV is not unusual, particularly where there may have been fatigue conditions leading up to the failure.

Microstructure It is usual to examine the microstructure of a failed component. This is a visual technique which only allows a qualitative assessment. A piece of the fracture surface known as a 'macro' is mounted in plastic and polished to a fine flat finish. The surface is examined under an optical microscope at ×100 and ×200 magnification and observations made about the condition of the material's microstructure. Microstructures vary widely, depending on the material specification and heat treatment. For carbon steels, ferrite and pearlite generally give desirable properties of tensile strength and toughness, whilst cementite and martensite cause a material to be hard and brittle. At lower magnifications, the fracture surface of a failed component may exhibit characteristic crescent-shaped 'beach marks', which are evidence of the operation of a fatigue mechanism.

Examination of the fracture surface of the crane sheave hanger showed some evidence of fatigue, propagated from the outer radius of the bend. The microstructure had acceptable levels of ferrite and pearlite with some coarsening of the grain structure visible in work-hardened areas around the outer radius of the bend.

Conclusions and recommendations

The findings so far are quite typical for a simple mechanical failure event. The results seem clear – but what can we conclude?

Visible evidence first. There is clear indication of a fatigue mechanism having been in operation, affecting first the outer radius of the crane sheave hanger bend. The work-hardened and coarsened microstructure are significant findings but are mainly consequential – caused *by* the fatigue mechanism, rather than being a separate cause themselves. The operational review has proved useful because it has effectively *eliminated* operational factors from the cause of failure. Dynamic stress concentrations that will be caused by swinging or jerking of the load are well within the factors of safety built into the crane design. It is the design review that links all the threads of evidence together; we have shown that excessive bending stresses *are likely*, owing to the bending moments caused by radial clearances between the sheave hanger and its sheave pin. These bending stresses will almost certainly exceed the material's fatigue limit – and possibly also its yield stress (R_e) under some conditions. This is what caused the failure.

Figure 4.5 shows a simple design change which will help eliminate this type of failure. The sheave hanger is fitted with a steel tube welded to its inside radius. The sheave pin is a good 'running fit' in the tube, thus eliminating the variation in radii of the contacting surfaces that caused the imposed large bending moments. The addition of the tube is a practical step – an equally good solution would be simply to bend the sheave bracket to the same radius as the sheave pin so it sits snugly over it. Practically, this is difficult – it is not easy to achieve close radial tolerances on the inside radius of bends, so the addition of the tube is a better solution.

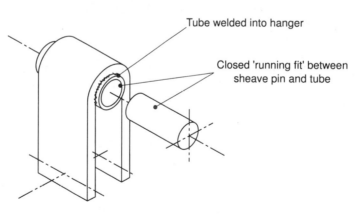

Tube welded into hanger

Closed 'running fit' between sheave pin and tube

Figure 4.5 A simple design change

4.4 Case study task

This case study task consists of five questions about engineering failures. They cover important general points about the *format* that engineering failures often take. Try to make your answers as brief as possible – long explanations are not required.

Q1: What kind of views and early conclusions can you expect to hear immediately after a failure occurs?

Q2: What are the five essential parts of an engineering failure investigation?

Q3: Which is the most important loading condition to look at when trying to decide why a component failed?

Q4: What are the three main objectives when performing examination of failed components?

Q5: What is the practicality of drawing firm conclusions in failure investigation cases?

References

4.1 BS 466 (1984). *Specification for power driven overhead travelling cranes.* British Standards Institution, London. This is an equivalent standard to ISO 4301/1.

5 Crane sheave – simple costing

Keywords

Costing and value engineering – costing breakdown – costs of design changes – cost–benefit analysis.

5.1 Objectives

The main objective of this case study is to look at the subject of simple costing and how it fits into the design process. The previous case study ended by recommending changes to the crane sheave hanger design. This may involve extra costs, so it is important to have some way of weighing the benefits if the design process is to run smoothly. Value engineering, often referred to also as value analysis (they mean the same), is a technique used to reduce costs *before they occur* during the detailed design or manufacturing stage. It also considers the quality and reliability of the component, so it is more than a blunt exercise in cost-cutting.

The technique of value engineering is not a separate discipline, as such; it is best thought of as an element of the design process – as part of it. It has links, therefore, with the processes of detailed design and the most effective implementation of value engineering is done by the designer, rather than a separate 'Value Engineer'. It is not easy, or particularly productive, to identify separate chronological steps; it is better to think of value engineering as a questioning attitude, applied throughout the design process. The questions to be asked are:

- Can this design be changed to give:
  - a saving in weight?
  - less manufacturing time?
  - lower material and/or labour costs?

The answers to these questions are sometimes arrived at intuitively, but for a more tangible output the results are often expressed as a *cost–benefit analysis*. This is a way of weighing up and comparing the effects of cost reductions, but taking a longer term, more enlightened view, sometimes involving probabilities. Although difficult, this is the essence of value engineering – the meticulous and persistent search for zero-cost changes that will improve an engineering product.

5.2 The problem

| Design Manager: | 'Fit a tube where?' |
| Student Designer: | 'In the inside of the sheave hanger.' |

'It doesn't need a tube, it's strong enough as it is.'
'It's not a case of strength, it's . . .'
'And there's plenty of safety built in; that sheave pin is at least twice as thick as it needs to be.'
'It's more a case of . . .'
'Case of what?, hold on' (telephone is ringing)
'Bending moments?'
'Plenty of bending resistance as well – look, this idea seems to be all costs – a fitted pin and tube; precision means costs, accuracy means costs and welding means costs, and then there's all the other costs of design changes: drawings, spares manuals, QA, tooling, work instructions . . . the list goes on . . .' (answers telephone)
'Hello George, yes, hold the line, what's the problem anyway?'
'Well, one's failed.'
'One, and how many have we made, 7000 or so?'
'One that we know about, and we seem to have been selling a lot of spare sheave hangers recently.'
'OK, bring me a proper proposal, my door is always open.'
'I can do that.'
'Yes, hello George . . . (closing door) . . . yes, about that golf weekend.'

5.3 Engineering costing

It is true that design changes often involve extra costs. Part of a good design process, however, is to quantify these costs – experienced people do develop an intuitive feel for cost implications, but it is wise to use such intuition to *help* with a proper quantification of cost changes, not to replace it. Remember also that calculating costs is only part of value engineering – it does not look at what other design or operational benefits might be achieved.

The manufacturing cost of an engineering component is divided into three categories: material costs, labour costs and overheads. The relative proportions of these vary, depending on the complexity of the design, materials, physical size, etc., so it is difficult to generalise. An assembly such as the crane sheave is made using batch-production methods, rather than the mass-production techniques that are used for consumer goods or motor vehicles. For such products the cost categories that are used are not 'fixed'; they will vary from product to product, depending on the actual techniques used by the manufacturer. Figure 5.1 shows a typical costing breakdown used for the crane sheave hanger. There are a few important points to note about the content of this breakdown:

- It makes several assumptions about the speed and efficiency with which manufacturing activities are done. Obviously, such assumptions should be kept constant throughout a costing analysis to keep the results in balance.
- Although the objective of a costing analysis is to predict the absolute cost of manufacturing a certain design, there are always assumptions and uncertainties

that limit the accuracy. A lot of the 'value' of cost analysis, therefore, is in the way that it can make relative comparisons between different designs or manufacturing methods.

- It excludes any consideration of overheads. Overheads are more a function of company policy than a design issue.
- For mass-production manufacturing it is possible to reduce significantly the errors and uncertainties of a costing analysis because the larger number of products made gives better 'proveability' of costs. It is not quite so easy for batch-manufactured components or individual 'one-off' designs.

5.4 Case study task

Your task is to write a short value engineering/cost report about the changes to the crane sheave hanger/pin fixing design proposed at the end of the previous case study. The report should comprise two sections:

- A revised *cost analysis*, based on the format of Fig. 5.1, but showing the cost implications of the new proposed design feature. You should estimate any new costs based on those apportioned to similar areas.

MATERIALS COSTS

Component	Material source	Unit cost	Cost
Sheave hanger	L.C.S 'flat' stock 80 mm × 10 mm	£5/metre	£2.50
Sheave pin	L.C.S bar stock 45 mm dia	£7/metre	£0.70
		Total	**£3.20**

LABOUR COSTS

Component	Activity	Time (mins)	Unit labour cost/min £	Cost £
Sheave hanger	Cut to length	2	0.3	0.6
	Heat and bend	12	0.3	3.6
	Mark out	7	0.3	2.1
	Drill in jig	6	0.3	1.8
	Check dimensions	5	0.3	1.5
Sheave pin	Cut to length	2	0.35	0.7
	Mount in lathe	3	0.35	1.05
	Face ends	3	0.35	1.05
	Cut threads	8	0.35	2.8
	Check dimensions	5	0.35	1.75
	Totals	**53**		**£16.95**

Total unit cost = £3.20 + £16.95 = £20.15 (1997 estimated prices)

Figure 5.1 Crane sheave – materials and labour costs

- A brief *cost–benefit analysis*, weighing the benefits of the proposed design change against its cost implications. Qualitative explanations are fine but you should still make clear reference to costs. Ideally, you should try to demonstrate how the revised sheave tube/pin assembly will be a nil-cost improvement. To do this it may be necessary to mention product quality and long-term reliability.

6 Casting machine – basic design

Keywords

The techniques of basic engineering design – understanding the approach. The anode casting machine mechanism – datums and tolerances – hole tolerancing (straightness, parallelism and squareness) – using BS 308 – shafts and bearings – limits and fits – using BS EN 20286 – specifying surface texture – allocating general tolerances.

6.1 Objectives

It is easy to be a little too eager to become involved in the innovative aspects of engineering design. This is fine, innovation is one of the cornerstones of the design process – it is also interesting and challenging. There is, however, an important step between the points at which the innovative stages of design end and manufacturing begins – this can be loosely termed *basic engineering design*. Figure 6.1 shows the content of each of these steps. Basic engineering design involves the use of published technical standards. There are thousands of these, referring to a bewildering array of engineering equipment; the most important ones however are those that deal with the very basic aspects of engineering design. These basic aspects are 'common denominators' to the way that all mechanical equipment is designed and made. These basic engineering standards are not new but perform an invaluable role because they represent proven technical experience.

The objective of this case study is to introduce some of these basic engineering aspects, and the technical standards that control them. So what are they? The main ones are related to two areas: those parts of an engineering assembly that *move*, i.e. they have movement relative to another part, and those that control the way that component parts of an engineering assembly *fit together*. Note that they do not, strictly, cover the way that components are loaded; loads and stresses are worked out during the earlier design stages. Taken together, the basic engineering standards divide into four subject-groups:

- basic dimensional accuracy
- tolerances
- limits and fits
- surface finish.

The activity of basic engineering design is about *applying* these established engineering practices (they are described by technical standards, remember) to a design concept, hence providing a bridge to the manufacturing activity. It is

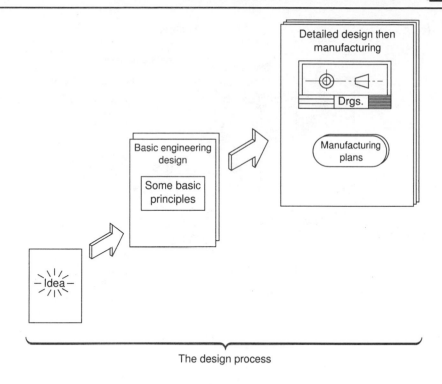

The design process

Figure 6.1 Basic engineering design

important to appreciate how these design activities work. It is worth emphasising two key points:

BASIC ENGINEERING DESIGN IS RARELY INNOVATIVE

but

IT IS AN ESSENTIAL PART OF THE DESIGN PROCESS

6.2 The casting machine mechanism

Figure 6.2 shows a new design of anode casting machine used in the manufacture of aluminium. An insulated ladle is filled with 500 kg of cast iron pieces which are melted by an electric heater. The molten cast iron is then poured into moulds to make cylindrical cast iron anodes, used in the electrolytic process to extract pure aluminium from its ore. The ladle is tilted by an electric motor-driven screw jack acting via a double cantilever mechanism. The mechanism is designed so that it lifts and tilts the ladle whilst maintaining a predetermined positioning of the ladle lip for accurate pouring. This is essential as the mould annulus into which the molten metal must be poured is less than 50 mm wide. Each ladle of metal ('the charge') fills about 15 moulds. The ladle and cantilever assembly is mounted on a fabricated steel frame which can slide in two planes on low friction ball-bearing tracks. Electric motors controlled by a computerised system are used to position the unit as required.

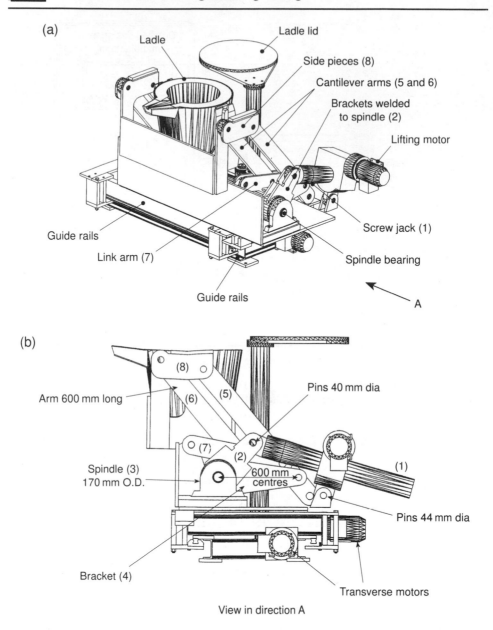

(a)

Ladle

Ladle lid

Side pieces (8)

Cantilever arms (5 and 6)

Brackets welded to spindle (2)

Lifting motor

Screw jack (1)

Guide rails

Link arm (7)

Spindle bearing

Guide rails

A

(b)

(8)

Arm 600 mm long

(6) (5)

Pins 40 mm dia

Spindle (3) 170 mm O.D.

(7)

(2)

600 mm centres

(1)

Bracket (4)

Pins 44 mm dia

Transverse motors

View in direction A

Figure 6.2 The casting machine

The casting machine works in a hot and dirty environment and operates continuously for 12 months between maintenance periods. Reliability is essential – if the machine breaks down, the entire aluminium-manufacturing production line has to stop. For this reason, the design of the moving parts is very robust; with the exception of the motors and gearbox, all the parts are slow-moving with large bearings and fixings. The cantilever arrangement is made of thick steel sheet to avoid twisting and deflections.

6.3 The problem

The problem is that the basic engineering design of the machine must be specified *properly* if it is to have the necessary reliability. The most important part is the cantilever mechanism as this cannot be maintained without shutting down the machine, and it must work absolutely correctly in order to lift the ladle smoothly and position it accurately during each working cycle. In practice, the cantilever mechanism and its associated steelwork forms the core of the casting machine design, other parts such as the motors and gearbox, ladle and bearing tracks being proprietary 'catalogue' items bought from component suppliers. The basic engineering design must incorporate proven principles of dimensional accuracy, tolerances, limits, fits and surface finish, within its moving and fitted parts. As well as ensuring that the machine will operate correctly this will also help with the specification of *manufacturing* information, needed by manufacturers so that they can prepare working drawings, cost estimates and production plans. This is an activity common to all engineering designs containing mechanical components, whether on a small or large scale.

6.4 Design details

Figure 6.2 shows the plate steel components that make up the cantilever mechanism. The spindle is also shown, fabricated from a seamed steel tube with plate end-caps and brackets. The various pins, bushes, bearings and fixings that complete the assembly are not shown in detail but nominal dimensions are given (note this terminology carefully: these are *nominal* dimensions, rather than accurate ones).

Operation

The mechanism operates as follows (refer to Fig. 6.2): the screw jack (1), driven by the motor and gearbox, rotates the spindle (3) via the clevis pin in the double brackets (2) which are welded to the machine spindle. This spindle is held in roller bearings, themselves mounted in cast-iron housings which are dowelled and bolted to the bedplate. The spindle movement is transferred to the cantilever arm arrangement by a further bracket (4), set at an angle to obtain the maximum lifting force. The cantilever comprises two parallel arms (5) and (6) which act via pins onto the side pieces (8). A link arm (7) connects the parallel arms and is held at one end, by a pivot pin, to the machine frame. There is a symmetrical set of linkages on each side of the ladle. In use, the linkages move quite slowly: the ladle takes about 45 seconds to move to the 'full tipping' position, and a similar time to return. The arms are connected by stainless steel pivot pins which run in phosphor-bronze bushes fitted into holes in the arms. These pins have a secure fixing so they cannot fall out in use. The phosphor-bronze bushes are impregnated with graphite lubricant and do not need greasing or other further lubrication throughout their life. The successful operation of the parts of the machine (their correct name is *machine elements*) is governed by some basic engineering design principles. We can look at these in turn.

Datums and tolerances

These are the two most fundamental principles; all machine mechanisms rely on them. A *datum* is a reference point or surface from which all other dimensions of a component are taken; these other dimensions are said to be *referred to* the datum. In most practical designs, a datum surface is normally used, this generally being one of the surfaces of the machine element itself rather than an 'imaginary' surface. This means that the datum surface normally plays some important part in the operation of the elements – it is usually machined and may be a mating surface or a locating face between elements, or similar. Simple machine mechanisms do not *always* need datums; it depends on what the elements do and how complicated the mechanism assembly is. Figure 6.3 shows the principle of a datum surface applied to an array of holes in a steel plate.

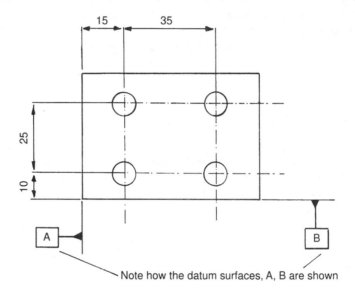

Note how the datum surfaces, A, B are shown

Figure 6.3 Datum surfaces

The second principle is that of dimension *tolerances*. The concept of a linear or angular dimension which is perfect exists only in theory – practically, a dimension will always have a variation about its 'perfect' value. It is important to look at how dimensional tolerances are expressed on engineering sketches and drawings. A dimension expressed as a single figure, 50 mm for example, refers only to the *nominal* size of that measurement – you may also see it referred to as the 'basic' size, the meaning is much the same. To specify it properly, to avoid misinterpretation, it needs to be toleranced. While this would be straightforward for our 50 mm linear dimension (we could specify it as 50 mm ± 0.1 mm, for example), this simple answer would not necessarily be applicable to the other types of dimensions found in machine elements: angles, flatness, parallelism, etc. Fortunately there is a well-developed set of standards covering this: British Standard BS 308 (ref. 1) contains several parts covering all of the dimensional possibilities that you need to know.

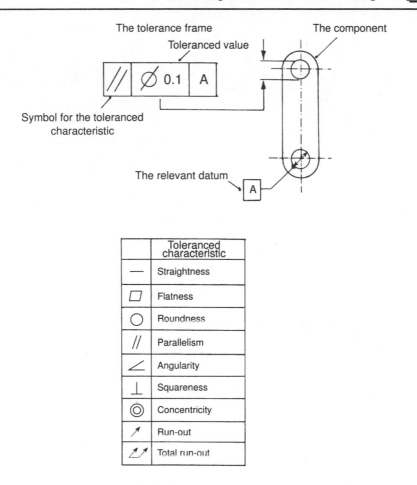

Figure 6.4 Symbols for toleranced characteristics

Figure 6.4 shows the main symbols that are used for expressing toleranced characteristics – note how the *tolerance frame* is used to display a set of tolerance information in a concise form. This outline description of datums and tolerances is very general; we will look in a little more detail at the principles of how they are used in the design of some machine elements.

Holes

The tolerancing of holes depends on whether they are made in thin sheet (up to about 3 mm thick) or in thicker plate material. In thin material, only two toleranced dimensions are required:

- *size*: a toleranced diameter of the hole, showing the maximum and minimum allowable dimensions;
- *position*: position can be located with reference to a datum and/or its spacing from an adjacent hole. Holes are generally spaced by reference to their centres.

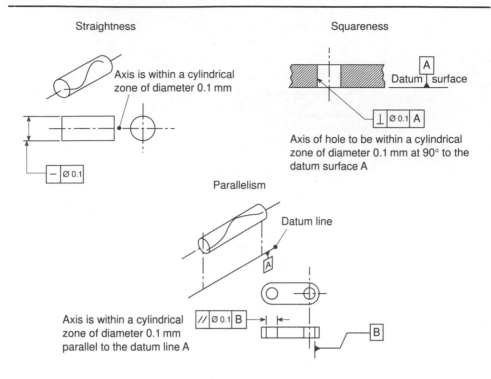

Figure 6.5 Straightness, parallelism and squareness

For thicker material, three further toleranced dimensions become relevant: straightness, parallelism and squareness (the symbols are shown in Fig. 6.4). These three *can* be confusing, but they are measuring fundamentally different things – Fig. 6.5 shows them in simplified form. Note the following key points:

- *straightness*: a hole or shaft can be *straight* without being perpendicular to the surface of the material;
- *parallelism*: this is particularly relevant to holes and is important when there is a mating hole-to-shaft fit;
- *squareness*: the formal term for this is perpendicularity. Simplistically, it refers to the squareness of the axis of a hole to the datum surface of the material through which the hole is made.

Shaft concentricity

Shaft concentricity is important whenever shafts or spindles connected to other machine elements run in bearings and are therefore *constrained by* these bearings. Errors in concentricity can cause undesirable stresses. Strictly, the concentricity of the surface of a circular shaft is referred to a datum axis, i.e. to show whether the real axis of the shaft is co-axial with a datum axis. Practically, the definition is often loosened to refer to the concentricity of a shaft axis with that of a smaller diameter part of the same shaft. The accepted symbol is shown in Fig. 6.4 and you can see it being used later in this chapter.

Angles

Tolerancing of angles is straightforward, using the tolerance frame symbols shown in Fig. 6.4. An angle is referred to a datum line or surface and the angular tolerance expressed using two parallel 'plane limits' rather than a plus/minus tolerance in degrees. Treat this as purely convention (which it is).

Bearings

Bearings are commonly used machine elements that permit free rotation of a shaft or spindle. Most bearings are designed to allow continuous rotation of the shaft through a full 360° but there are also applications where the rotation is partial, or intermittent. There are numerous different makes of bearings but they generally fall into one of the following two types.

Journal bearings   In a journal bearing, the shaft rotates inside a loose-fitting bearing shell of softer, often porous, bearing material. Lubricant, such as oil, grease or a low-friction compound like PTFE or graphite is used between the surfaces. The shell is sometimes split into two halves. The bearing shell is fitted tightly into its static housing to stop it revolving with the shaft. For simple rotation applications, journal bearings are designed to be parallel within close limits (about 0.1 mm), thereby avoiding excessive loads. A simple journal bearing is shown in Fig. 6.6.

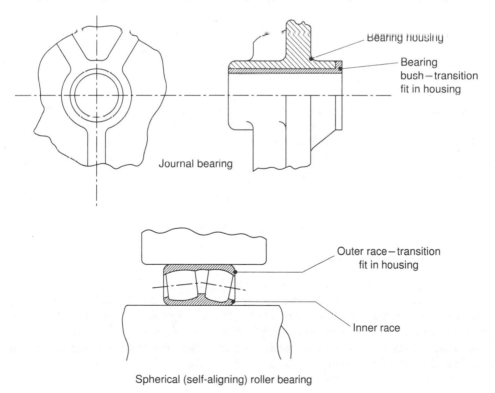

Bearing housing

Bearing bush – transition fit in housing

Journal bearing

Outer race – transition fit in housing

Inner race

Spherical (self-aligning) roller bearing

Figure 6.6 Bearing types

Rolling contact bearings This category contains both ball and roller-type bearings: elements designed to transfer loads between rotating and stationary machine elements and permit free rotation with a minimum of friction. Consisting of a rotating inner race, a set of balls (or rollers) and a stationary outer race, or 'track', they are machined to very fine tolerances. They may be of either parallel or tapered configuration – a common design has a slightly spherical outer race so that it will accommodate misalignment of up to 2.5°. This is of great practical use in large, robust machine elements where high accuracies are not always possible. The inner race needs to fit tightly on the rotating shaft so that it rotates with it without slipping. Similarly, the outer race must fit tightly into the stationary housing so that it does not rotate with the moving balls or rollers. Both fits must not be so tight as to cause distortion of either race, which will cause the bearing to fail rapidly in use. Rolling contact bearings are proprietary items available in a variety of preferred sizes.

Shaft/hole fits

In machine element design there is a variety of different ways in which a shaft and hole are required to fit together. Elements such as bearings, location pins, pegs, spindles and axles are typical examples. Depending on the application, the shaft may be required to be a tight fit in the hole, or to be looser, giving a clearance to allow easy removal or rotation. The system designed to establish a series of useful fits between shafts and holes is termed *limits and fits*. This involves a series of tolerance grades so that machine elements can be made with the correct degree of accuracy and be interchangeable with others of the same tolerance grade.

The British Standard BS 4500 / BS EN 20286 (ref. 2) 'ISO limits and fits' contains the recommended tolerances for a wide range of engineering requirements. Each tolerance grade is designated by a combination of letters and numbers, such as IT7, which would be referred to as grade 7. Figure 6.7 shows the principles of a shaft/hole fit. The 'zero line' indicates the basic or 'nominal' size of the hole and the shaft (it is the same for each) and the two shaded areas the tolerance zones within which the hole and shaft may vary. The hole is conventionally shown above the zero line. The algebraic difference between the basic size of a shaft or hole and its actual size is known as the *deviation*. Three points follow from this:

- it is the deviation that determines the nature of the fit between a hole and a shaft;
- if the deviation is small, the tolerance range will be near the basic size, giving a tight fit;
- conversely, a large deviation gives a loose fit.

Various grades of deviation are designated by letters, similar to the system of numbers used for the tolerance ranges. Shaft deviations are denoted by small letters and hole deviations by capital letters. Most general engineering uses a 'hole based' fit in which the larger part of the available tolerance is allocated to the hole (because it is more difficult to make an accurate hole) and then the shaft is made to suit, to achieve the desired fit. Figure 6.8 shows those fits most commonly used for machine elements. Seven main combinations are of interest:

- *Easy running fit*: H11–c11, H9–d10, H9–e9. These are used for bearings where a significant clearance is necessary.

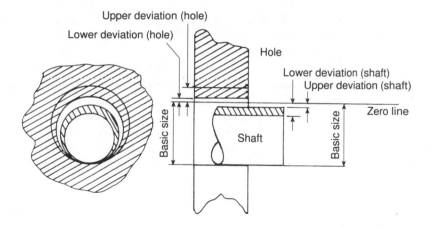

Figure 6.7 Principles of a shaft/hole fit

	Clearance fits					Transition fits		Interference fits	
Holes	H11	H9	H9	H8	H7	H7	H7	H7	H7
Shafts	c11	d10	e9	f7	g6 / h6	h6	k6 / n6	n6 / p6	p6 / s6

	Easy running		Close running		Sliding	Push	Drive	Light press	Press
Nominal size in mm	Tols* H11 c11	Tols H9 d10	Tols H9 e9	Tols H8 f7	Tols H7 g6	Tols H7 h6	Tols H7 k6	Tols H7 n6	Tols H7 p6 / H7 s6
6–10	+90 / 0 ; −80 / −170	+36 / 0 ; −40 / −98	+36 / 0 ; −25 / −61	+22 / 0 ; −12 / −28	+15 / 0 ; −5 / −14	+15 / 0 ; −9 / 0	+15 / 0 ; +10 / +1	+15 / 0 ; +19 / +10	+15 / 0 ; +24 / +15 ; +15 / 0 ; +32 / +23
10–18	+110 / 0 ; −95 / −205	+43 / 0 ; −50 / −120	+43 / 0 ; −32 / −75	+27 / 0 ; −16 / −34	+18 / 0 ; −6 / −17	+18 / 0 ; −11 / 0	+18 / 0 ; +12 / +1	+18 / 0 ; +23 / +12	+18 / 0 ; +29 / +18 ; +18 / 0 ; +39 / +28
18–30	+130 / 0 ; −110 / −240	+52 / 0 ; −69 / −149	+52 / 0 ; −40 / −92	+33 / 0 ; −20 / −41	+21 / 0 ; −7 / −20	+21 / 0 ; −13 / 0	+21 / 0 ; +15 / +2	+21 / 0 ; +28 / +15	+21 / 0 ; +35 / +22 ; +21 / 0 ; +48 / +35
30–40	+140 / 0 ; −120 / −280	+62 ; −80	+62 ; −50	+39 ; −25	+25 ; −9	+25 ; −16	+25 ; +18	+25 ; +33	+25 ; +42 ; +25 ; +59
40–50	+160 / 0 ; −130 / −290	0 ; −180	0 ; −112	0 ; −50	0 ; −25	0 ; 0	0 ; +2	0 ; +17	0 ; +26 ; 0 ; +43

* Tolerance units in 0.001 mm

Figure 6.8 Common shaft/hole fits

- *Close running fit*: H8–f7, H8–g6. This only allows a small clearance, suitable for sliding spigot fits and infrequently used journal bearings. This fit is not suitable for continuously rotating bearings.
- *Sliding fit*: H7–h6. Normally used as a locational fit, in which close-fitting items slide together. It incorporates a very small clearance and can still be freely assembled and disassembled.
- *Push fit*: H7–k6. This is a transition fit, mid-way between fits that have a guaranteed clearance and those where there is metal interference. It is used where accurate location is required, e.g. dowel and bearing inner-race fixings.
- *Drive fit*: H7–n6. This is a tighter grade of transition fit than the H7–k6. It gives a tight assembly fit where the hole and shaft may need to be pressed together.
- *Light press fit*: H7–p6. This is used where a hole and shaft need permanent, accurate assembly. The parts need pressing together but the fit is not so tight that it will overstress the hole bore.
- *Press fit*: H7–s6. This is the tightest practical fit for machine elements. Larger interference fits are possible but are only suitable for large heavy engineering components. The H7–s6 is used for tight-fitting components such as bearing bushes.

Surface finish

Surface finish, more correctly termed 'surface texture', is important for all machine elements that are produced by machining processes such as turning, grinding, shaping or honing. This applies to surfaces which are flat or cylindrical. Surface texture is covered by its own technical standard, BS 1134 *Assessment of surface texture* (ref. 3) whilst the relevant drawing symbols are also given in BS 308 mentioned previously. Surface texture can affect the way that mating or locating parts such as holes/shafts or plane surfaces fit together so it has important links with the accuracy of hole/shaft tolerance grades. It is measured using the parameter R_a which is a measurement of the average distance between the median line of the surface profile and its peaks and troughs, measured in micrometres (μm). There is another system from a comparable standard, DIN ISO 1302 (ref. 4) which uses a system of N-numbers – it is simply a different way of describing the same thing. Details are shown in Fig. 6.9. As a 'rule of thumb', a rough turned surface, with visible tool marks, is about grade N10 (12.5 μm R_a) and a reasonably smooth machine surface is likely to be about grade N8 (3.2 μm R_a). Surfaces which mate with other static surfaces, or provide a datum surface, will usually be specified as grade N7 (1.6 μm R_a) or better. Surfaces which incorporate a relative movement or bearing function vary from grade N6 (0.8 μm R_a) down to the finest 'normal' surface grade N1 (0.025 μm R_a). Finer finishes can be produced but are more suited for precision application such as instruments. It is good practice to specify the surface finish of close fitting surfaces of machine elements, as well as other BS 308 parameters such as squareness and parallelism.

General tolerances

One sound principle of engineering practice is that of general tolerances. In any machine design there will actually only be a small number of *toleranced features*

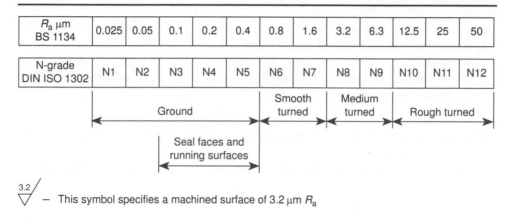

R_a μm BS 1134	0.025	0.05	0.1	0.2	0.4	0.8	1.6	3.2	6.3	12.5	25	50
N-grade DIN ISO 1302	N1	N2	N3	N4	N5	N6	N7	N8	N9	N10	N11	N12

Ground — Smooth turned — Medium turned — Rough turned

Seal faces and running surfaces

$\overset{3.2}{\triangledown}$ — This symbol specifies a machined surface of 3.2 μm R_a

Figure 6.9 Surface finish

(remember this terminology?) – the remainder of the dimensions will not be critical. There are two ways to deal with this: first, an engineering drawing or sketch can be annotated to specify that a *general tolerance* should apply to features where no specific tolerance is mentioned. This is often expressed as ±0.5 mm. Alternatively, the drawing can make reference to a 'general tolerance' standard such as BS EN 22768 (ref. 5) – it gives typical tolerances for linear dimensions as shown in Table 6.1.

Table 6.1

Dimension	Tolerance
0.6 mm – 6.0 mm	±0.1 mm
6 mm – 36 mm	±0.2 mm
36 mm – 120 mm	±0.3 mm
120 mm – 315 mm	±0.5 mm
315 mm – 1000 mm	±0.8 mm

It is easy, once you have become familiar with these detailed technical aspects, to lose sight of the overall objective of the basic engineering design phase of engineering design, and to want to apply these techniques of tolerancing and fine metrology to every part of a machine or mechanism. The skill lies in *selection*; good design is about taking pains to use these systems when they are needed but not to use them excessively, or to think that they are somehow a substitute for innovative design thinking.

6.5 Case study task

The objective of this case study task is to practise using the various aspects of basic engineering data that are needed to specify fully an engineering design. The anode casting machine uses a variety of common machine elements. For this case study we are concerned only with the cantilever-based mechanism used to tilt the ladle and return it to its original position. Taking the basic design as presented in Fig. 6.2 without changing the arrangement or dimensions, your task is to add to it that basic design information which is essential in order to *specify* properly the design. The components should be capable of being manufactured, so the necessary information will be needed concerning:

- datums and tolerances;
- data such as straightness, parallelism, squareness and concentricity of the key machine elements;
- bearings (there are both journal and rolling contact types in the mechanism);
- shaft/hole limits and fits;
- surface texture of mating and locating parts.

The easiest way to show all this information is on fully dimensioned sketches. The components are simple, so there is little point in doing full engineering drawings – all the necessary information can be put on a sketch. You can use the technical standards quoted where they are applicable, although most of the information needed is included in the case study text. Don't over-specify the components, but remember that one of the main design criteria of the casting machine mechanism is the need for it to be robust and reliable, because of the job it does.

Some advice on methodology

Chapter 2 looked at the concept of dividing design problems into different types, based on the combination of their technical and procedural characteristics, and their level of 'difficulty'. This problem is one of the simpler types, very close to the linear technical model shown in Fig. 2.3 earlier in this book. Note how the task is one of working through a series of well-defined technical steps. Although the casting machine mechanism is a straightforward example, it is a general rule that the generic activity of all basic engineering design (as defined at the beginning of this case study) follows this pattern. The activity is much the same whether a mechanism has ten, or a thousand, dimensions, tolerances or limits that need defining. So, to summarise the approach you should apply to this case study task:

- First, read Chapter 2, particularly the section about *linear technical problems*.
- Address basic design issues such as dimensional tolerances, surface texture, etc., separately, in turn, to avoid confusion. But make sure you consider them all.
- Use the information from *technical standards*. Accept that these contain well-proven information.
- Don't try to be too innovative when dealing with basic engineering design (you will have your opportunity in other case studies in this book).

References

6.1 BS 308: Part 2 (1992) *Recommendations for dimensioning and tolerancing of size*; and BS 308: Part 3 (1990) *Recommendations for geometrical tolerancing*. British Standards Institution, London.

6.2 BS 4500 (1990) *ISO limits and fits*. British Standards Institution, London. Part 1 Section 1.1 is equivalent to BS EN 20286–1 and Part 1 Section 1.2 is equivalent to BS EN 20286–2.

6.3 BS 1134: Part 1 (1988) *Assessment of surface texture – methods and instrumentation*. British Standards Institution, London. This is a similar standard to DIN/ISO 1302.

6.4 DIN ISO 1302 (1992) *Technical drawings – methods of indicating surface texture*. Berlin, Geneva.

6.5 BS EN 22768–1 (1993) *Tolerances for linear and angular dimensions without individual tolerance indications*; and BS EN 22768–2 (1993) *Geometrical tolerances for features without individual tolerance indications*. British Standards Institution, London.

The *Rainbow* sculpture – innovation in design

Keywords

Innovation – outside design influences – architects' descriptions. Improving conceptual design – using intuition. Embodiment design of structures; form, stress distribution, materials of construction and quality classification. Detailed engineering design – the loading regime – vibration frequencies – structural damping – fatigue resistance. Manufacturing influences; shape, cross-section and fabrication.

7.1 Objectives

Remember this statement made in Chapter 1?

INNOVATION IS THE BEDROCK OF THE DESIGN PROCESS

Fine – but where does all this innovation come from? It normally happens at the *beginning* of the design process, during the conceptual engineering phase, and may also carry forward into the embodiment design phase, and have its effect there. Sometimes, innovation comes from designers – there are legendary (and true) stories about engineering designers seeing a problem or opportunity and then providing a design to fit it. Many engineering designers are good, conceptual, innovative thinkers – and some are not. It is also common, however, for the innovative spark to come from outside, from others, the most common sources being architects, industrial engineers and 'creative designers', roughly in that order. These outside influences (who tend to become compressed into the general category of 'architects') share the common feature that they are several steps removed from the activities of detailed design and manufacture. Perhaps it is this freedom from mundane practical considerations that encourages the process of innovation. The objective of this chapter is to look at how innovation meshes with the practicalities of engineering design. In doing so, we will need to consider the role of these outside influences.

Outside influences – an example

Engineering sculptures provide a good example of the role of architects in engineering design. Innovative sculptures are becoming popular in both in-dustrialised and developing countries. They may have a practical value (as a functional bridge or tower of some sort) but most are purely ornamental – their only purpose is to look nice. The steel sculpture at the entrance to Sweden's Stockholm harbour was the subject of lively design discussions between the City authorities, architects and engineers. The successful firm of architects had summarised their idea, during a quayside presentation, like this:

> Our proposed sculpture will provide a *fulfilling backdrop* to this historic harbour. Replicating the concept of a *welcoming gateway*, it will demonstrate to visitors the city's *vibrant commercial potential*. The *pleasing curve* of the structure represents *grace, elegance and innovation*.

The harbour committee nodded vigorously in approval, following up with unanimous warm, but cautious, applause. Relishing the enthusiasm of the moment, the City Mayor followed on:

> The Committee looks forward to the commissioning of this fine sculpture – there is little doubt that it will be the epitome of *fine and exacting workmanship*, a credit to the engineers who will build it.

and then as an afterthought:

> ... of *natural* materials, in *harmony* with our surroundings ...

The final words (and strains of the band) fought with the swirling wind that seemed to be almost a part of Stockholm harbour.

7.2 Your problem – the *Rainbow* sculpture

There are often problems with designing something new. The *Rainbow* is no exception – there is the constraint that conceptual design 'limits' have already been set, and against the background of architects' requirements, the sculpture has to be designed using sound engineering principles and practices. It must have a long lifetime, be economical to manufacture, and be safe. The design process consists of the three established stages:

* conceptual design
* embodiment design
* detailed engineering design.

For a 'new' design, each of these stages contains challenges. In a practical design situation these challenges can be of two sorts: those caused by the *nature* of the engineering disciplines themselves and those which are there as a direct result of the *outside influences* that are imposed on the design. The problem is how to handle these – together – whilst still producing a good design. We can look at this in a little more depth, starting with the conceptual design phase.

Conceptual design

Figure 7.1 is a reproduction of the architects' concept sketch for the Stockholm harbour *Rainbow* sculpture. It is shown as a 180° parabolic arch spanning the harbour

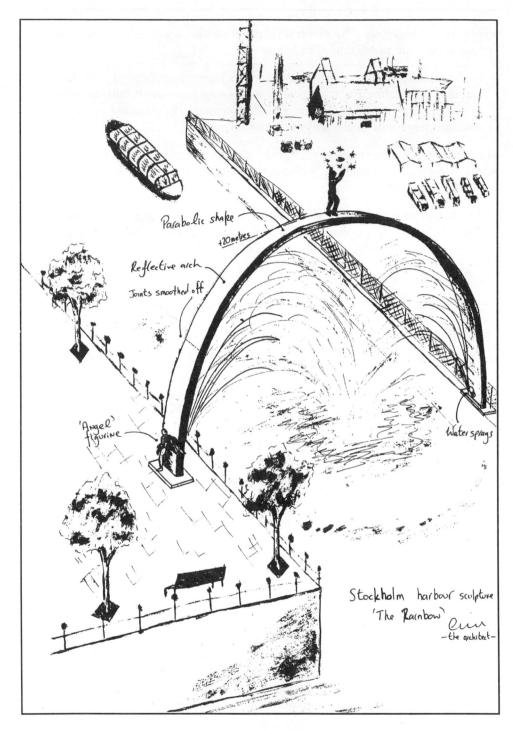

The following labels appear within the sketch:

- Parabolic shape
- +20 metres
- Reflective arch
- Joints smoothed off
- 'Angel' figurine
- Water sprays
- Stockholm harbour sculpture 'The Rainbow'
- —the architect—

Figure 7.1 The architect's concept sketch

entrance. A vertical clearance of 20 metres at middle span allows access for small commercial vessels and high-masted pleasure yachts. The sketch shows water-jets spraying from both abutments so the arch looks as if it is set in a fountain, surrounded by water spray mist. This is illuminated, at night, by powerful halogen lights. Two figurines stand on the arch; one is located at the end and the other stands on top of the arch at its highest point.

Embodiment design

A few embodiment design criteria are inferred from the architect's sketch. Overall arc dimensions are specified. The arch is described as reflective and possibly intended to be chromium plated. The cross-sectional shape is not specified although a heavy chunky section would probably be unacceptable. Outside these limits there is room for choice. Interestingly, there is no information on cost – which is not uncommon.

Detailed engineering design

Engineering design features are rarely shown in architects' sketches. It is clear, however, that the arch has no external stiffeners, it must 'stand alone' without external braces or wires. This has an important influence on the cross-sectional area and material thickness that must be used to provide the necessary strength – but these requirements are *inferred* rather than being stated explicitly. It is also inferred from the sketches that the arch can be fabricated in sections to make up the parabolic shape. This is actually a necessity – it would be almost impossible to manufacture the sculpture in one piece, unless it was solid (which would be prohibitive on both weight and cost).

7.3 The Rainbow: *design process*

One purpose of this case study is to show how innovative features are both decided and dealt with as part of the design process. The best way to do this is to work through the example step by step, looking at the main points. This example of the *Rainbow* sculpture is by no means unique – you can think of it as a general case, using *principles* that hold good for many different design cases.

The Rainbow: *conceptual design*

The innovation process normally starts at the stage of conceptual design. It is important to avoid a type of paradox that can arise here: just because conceptual design *limits* have been set by an architect or other outside influence does not mean that further conceptual improvement is impossible. It is the designers' job to look for better ways to *express* the concept. There is often a better solution waiting to be found – so why not look for it? The most effective way of finding a concept solution

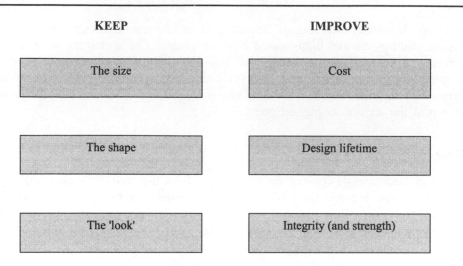

Figure 7.2 Improving conceptual design

is, frankly, based largely on intuition. Figure 7.2 shows a useful model that can help this intuition. It relies on the general observation that conceptual design solutions can be thought of as being generated from a 'stand-off' between two sets of influences: the basic objective is to keep to the required size, shape and 'look' of the design, whilst improving upon cost, lifetime and integrity. These influences form a good 'test' for conceptual improvements of any design.

Figure 7.3 shows one way to apply this idea to the *Rainbow* sculpture. The approach shown in Fig. 7.3(a) is a common one, an attempt to reduce the structural dimensions to save material. Although the main purpose of the *Rainbow* is aesthetic, it must still be strong enough to resist its own weight and imposed wind/snow loadings – so there is a limit on how far the structure's dimensions can be reduced. Excessive reductions will reduce the design factors of safety and compromise the integrity of the structure. There is often little chance for real conceptual improvement like this. A better solution is offered by Fig. 7.3(b). Here, the full parabolic arch is replaced by a half arch extending from a single abutment to the mid-span position. The other half of the parabola is replaced by a water jet, which, under the influence of gravity, forms a parabolic shape. The full arch 'gateway' effect is kept, as is the water spray and lighting display, albeit in a slightly different format. The main advantage of this concept is that the material requirement is reduced, giving a cost saving. There are, however, embodiment and detailed design implications of this approach, such as:

- possible increased bending moments owing to the single-point mounting;
- heavier single foundation required;
- larger deflections of the finished structure;
- a stiffer cross-section may be required.

It is important to *identify* such practical design considerations at this early stage so you can start to consider them properly during the subsequent embodiment design stage.

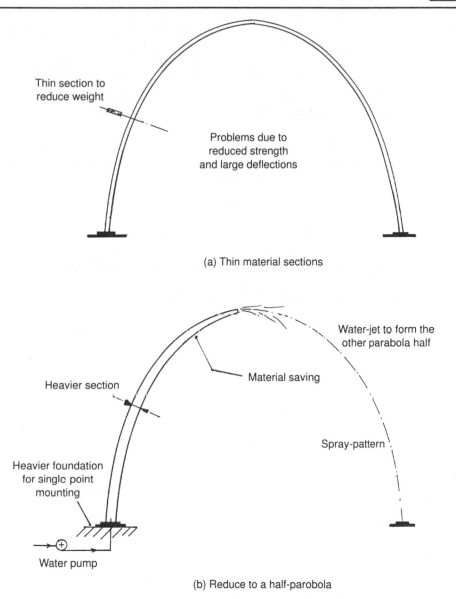

(a) Thin material sections

(b) Reduce to a half-parobola

Figure 7.3 Conceptual design solutions

The Rainbow: *embodiment design*

The process of embodiment design is, by definition, innovative. The action of looking for and evaluating different 'design ways' of doing things encourages the process of innovation, so the problem tends to become one of *controlling* the innovation rather than having to encourage it. It is often difficult to decide precisely the point at which embodiment design becomes detailed engineering design. The *Rainbow* is a good example of this – the simple structural shape means that aspects such as the form and structure of the design (traditionally embodiment issues) are closely linked to the stress and strength calculation which are generally considered as belonging to 'detailed design'. Don't worry about this too much, it is the action of *considering* each design area thoroughly that is important; the terminology is only there to help bring order to the activities, not replace their rationale. There are a few useful guidelines you can follow; the following five topics are *generally* considered embodiment issues – but there are no hard and fast rules on this.

Form   The shapes of complete assemblies and their larger subcomponents are often referred to, in design terminology, under the generic title of *form*. For the *Rainbow* the main form considerations are:

- the slenderness ratio of the 'beam' making up the main structural arc;
- the cross-sectional shape and area of the beam – chosen mainly for aesthetic reasons but also for its resistance to bending moment and shear force.

Stress distribution   This is to do with the general principles of the way that the stress is distributed throughout the structure (rather than calculation of precise numerical values). For the *Rainbow* the best method is to aim for an almost uniform stress distribution over the length of the parabolic beam. Note that this refers to that stress which acts as the *constraint* to the design – in practice this may be pure tensile stress (σ_t), or shear stress (τ), depending on which section is considered. Figure 7.4 shows how this is achieved – the lower end of the arc where bending moments and resultant shear forces are higher is designed to be 'stiffer' by using a larger cross-sectional area and thicker material.

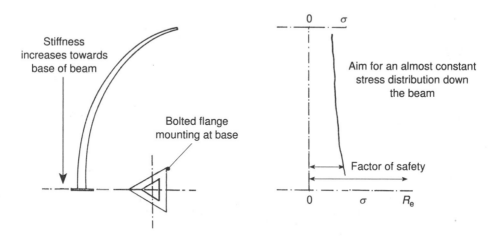

Figure 7.4 Embodiment design – stress distribution

Materials of construction   The basic material type is normally decided at the embodiment design stage – decisions about specific material grades, finishing and surface treatment are left until later, along with ways that the material can be cut, prepared and joined. The best material for the *Rainbow* is stainless steel. Its main features are:

- good corrosion resistance in marine atmosphere;
- it can be shotblasted and polished to a smooth reflective finish;
- it is weldable (depending on the grade chosen);
- strength – stainless steel has a yield strength (R_e) approximately 30–40 per cent higher than low carbon structural steels.

A chromium plated low carbon steel could be used but this is an expensive solution and would cause difficulties with structural welding and corrosion. A silver painted finish is another alternative; some aluminium-based paints can give a good finish, although are best applied in controlled factory conditions rather than after assembly of the structure on site. On balance, stainless steel is the best choice.

Ground fixings   Embodiment design options for steel civil structures normally reduce to the question of whether the end of the structure is secured by embedding in concrete or by bolting to a plinth. The answer depends on the weight and imposed loads on the structure – the limiting factor for a bolted fixing is the tensile yield and shear strength of the fixing bolts and the flange. If there is any uncertainty about the loading regime, then it is normal to use concrete fixings for safety. The *Rainbow* uses bolted fixings (see Fig. 7.4), which are strong enough to resist the calculated loads and allow easy assembly.

Quality classification   This, at best, is a rather intangible aspect of embodiment design. It cannot be ignored, however, and it is specifically included in published technical standards covering loaded structures such as British Standard BS 8100 (ref 1). The basic idea is that a structure is designed and built to a *quality class*, A, B or C. The final class depends on subratings achieved by the structure in five areas:

- design code
- design checking and testing
- materials' compliance with codes and standards
- workmanship and inspection during manufacture
- inspection and maintenance in use.

Figure 7.5 shows the derivation of the 'Class A' quality classification used for the *Rainbow* steel sculpture. Yes, this is a rather crude method – it cannot compete, for instance, with the detailed reliability analyses used for rotating machinery. It is, however, adequate for the application, and is in common use. The *Rainbow*, as it is a structural item, needs at least a 'B' quality classification. In practice, as it is made of stainless steel which needs proper attention to welding procedures, etc., it would be possible to upgrade this to an 'A', without much extra cost or effort.

The Rainbow: *detailed engineering design*

Detailed design follows on from the engineering decisions made at the embodiment stage and includes both qualitative and quantitative (calculation) activities. In terms of methodology, the detailed design phase involves *looking for problems to solve*. The problem-solving techniques themselves are well defined, particularly for structural engineering components, and have a robust theoretical origin. As with most aspects of design, an understanding of how materials and components fail is necessary. It would be wrong to suggest that all detailed engineering design activities

Activity	Quality 'consideration'	Rating
1. Design	• Full code compliance	2
	• Designed to 'code intent', with some simplifications	1
	• Simplified design method only	0
2. Design appraisal	• Full type-testing	2
	• Independent design appraisal	1
	• No independent design check	0
3. Materials of construction	• Full compliance with published material standards	1
	• Non-standard material, or manufacturer's untested own-brand materials	0
4. Manufacturing inspection	• Inspection as specified by the applicable code or construction standard	2
	• Random or unspecified inspection	1
	• No inspection	0
5. In-service inspection and maintenance	• Regular scheduled inspection and repair of defects/problems	2
	• Selective inspection and maintenance (as required)	1
	• Visual inspection only – no real maintenance	0

Deciding the quality classification

As a general guideline:

Structure class	Rating total
A	≥ 7
B	≥ 4
C	< 4

Figure 7.5 Quality classification – a guideline

are similar, or follow predictable paths; so much is design or discipline-specific. For engineering structures such as towers, bridges, masts and some lattice-type buildings, however, a broad pattern emerges: detailed design activities can be subdivided into two categories:

• design for strength (so it remains standing);
• design for manufacture (so it is possible to make it).

We can look at these and how they apply to the detailed design of the *Rainbow* (see Fig. 7.6).

The loading regime This is a 'design for strength' issue. Figure 7.7 shows the main loads that we can anticipate will act on the structure. There are three: self-weight (often known as 'dead-load'), wind-loading and snow/ice loading. Of these, the most important for the *Rainbow* is wind-load. Wind-loading is treated as having both a static component (constantly blowing wind) and a dynamic effect (caused by gusts), giving effectively two different, superimposed, loading conditions. The constant wind force gives a UDL (uniformly distributed load) up the *plane* of the beam, which tends to cause bending. The secondary effect of this – the crosswind component caused by the well-known 'vortex-shedding' phenomenon – acts

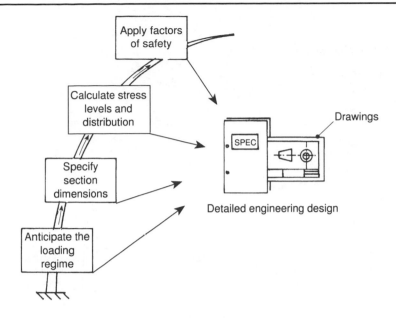

Figure 7.6 Detailed design – typical steps

perpendicular to this, trying to bend the beam in the other (crosswind) plane. This crosswind force tends to be cyclic (and rather unpredictable) and so is treated as having the same effect as wind *gusts*. Average wind speeds in harbour areas are high, so a figure of 20–25 metres/second is a suitable design figure to use. More accurate numbers based on geographical and terrain factors are available from design codes such as BS 8100. The direction of the wind load is also important – it has to be treated as unpredictable and the orientation of the beam cross-section chosen with this in mind. Snow and ice-loading of the beam can be simply superimposed on top of the beam's self-weight as shown in Fig. 7.7. The values (indicated by the length of the force arrows) are 'order of magnitude' only, but show clearly that dead loads are secondary to the wind loads in this type of design.

One aspect which is important to look at, particularly in structures, is *vibration*. Vibration is caused predominantly by wind loads, specifically the gust effect turbulence that is induced in the crosswind direction as previously described. The most dangerous condition is when the frequency of the induced vibration approaches the resonant frequency of the structure, producing increasingly large deflections and failure. The predominant natural frequency of the *Rainbow* is calculated using a finite element method and is about 1–2 Hz. The design parameters of the *Rainbow* are influenced by the need to limit stresses at the lower end. An increase in plate thickness in this area will increase the stiffness of the structure, thereby increasing the natural resonant frequency – this would lead to greater wind load effects (worse vibration). To prevent this, a heavy lead mass is added to the top of the structure – this reduces the natural resonant frequency, counteracting the effects of the stiffening at the lower end (see Fig. 7.8). The general term for this is 'structural damping'.

The beam cross-section   The cross-sectional shape that can be used for the beam of the *Rainbow* is constrained by its aesthetic requirement – it is one of the

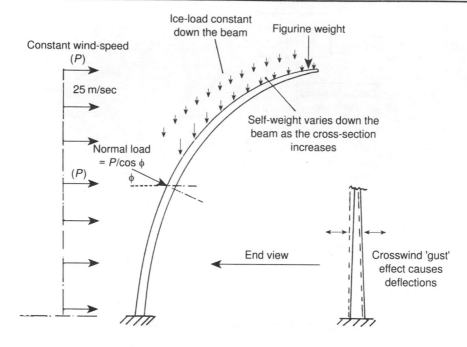

Figure 7.7 The beam loading regime

architects' ideas that it must be sleek and 'look good'. From a practical design viewpoint, the main purpose of the beam cross-section is to resist bending due to the wind loads, with the additional problem of the wind direction being variable. The general bending case of the beam is that of 'bending of curved bars', a well-established theory in which the initial curvature of a beam causes higher stress levels remote from the beam's neutral axis than for a beam which is straight. For the *Rainbow*, however (see Fig. 7.9), it is clear that the minimum L/D (beam length divided by cross-section depth) is about 9. This is classed as only a 'slightly curved' beam (i.e. when $L/D > 5$) hence the initial curvature has not much 'stress increasing' effect. This means that the beam cross-section does not have to be 'specially chosen' to allow for the parabolic shape of the beam.

Figure 7.9 shows some possible sectional shapes. The main influence on choice is the variable direction of the wind. If the (a) rectangular or (b) semi-elliptical section is used, the beam will be much stiffer in the lateral (crosswind direction) plane than in the perpendicular plane. This leads to a design which is 'uneconomical', in which material is wasted. Conversely, making the cross-section square would not really fulfil the aesthetic requirements. The triangular section (c) has several advantages. Being equilateral, it presents a similar profile in several wind directions – this also applies to the crosswind direction which is affected by the vortex shedding 'gusts'. A triangle also allows the 'full side' to be positioned on the outside curvature of the beam bow, which is the location of maximum tensile stress – the triangle apex, located at the inside surface of the bow, is less important. The neutral axis of the section is predictable, using simple geometry, so the section modulus and moment of area (and the resultant tensile and compressive stresses) can be easily calculated. Figure 7.9 also shows how the section size increases towards the base of the beam,

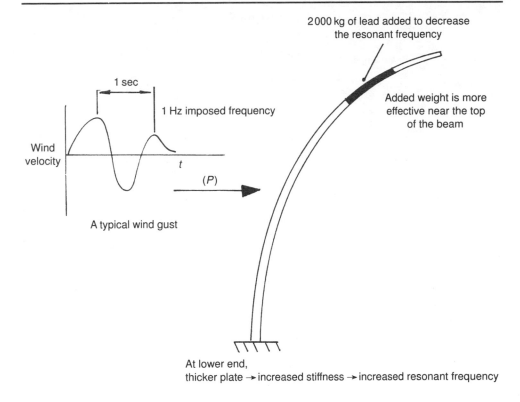

Figure 7.8 Beam vibration and damping

as the magnitude of the bending moment increases. This again is good use of the principles of *economic design* – because material is not wasted.

Deflections   The amount by which a steel structure deflects under load is calculated as part of the detailed stress calculations. There are well-known techniques, given in nearly all textbooks, dealing with deformable steel structures. There is also a less tangible relative to deflection – the criterion of *serviceability*. This is mentioned in some technical standards. A general guideline is that the best serviceability class is that achieved when the structure does not experience deflections of greater than 50 per cent of its limiting deflection over its lifetime. Serviceability is a function, therefore, of the flexibility of the beam, in turn related to its cross-section and material thickness.

Fatigue resistance   Resistance to fatigue is perhaps *the* most important detailed design criterion – it is firmly related to the 'design for strength' category identified earlier. The *fatigue limit* for a component can be much lower than the specified nominal yield strength (R_e) and is heavily influenced by the detailed design of the component. Features such as rounded corners and smooth transitions are used specifically to avoid lowering the fatigue limit too far. There is nothing unique about any of these features – they are general good practice for all structural steelwork designs.

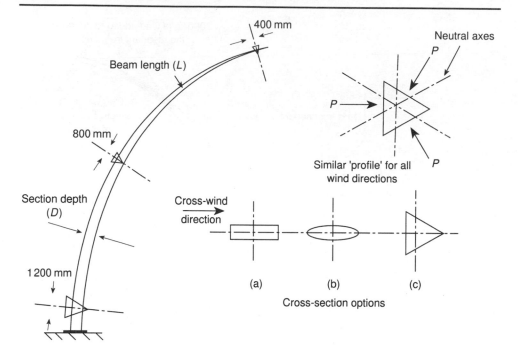

Figure 7.9 Beam cross-section shape

Factors of safety   Factors of safety are a fact of life in detailed engineering design. Their purpose is to take account of uncertainties that exist in the whole process of engineering design and manufacture. These uncertainties exist in several areas:

- *Loadings*: the loading regime on a structure is difficult to quantify exactly. Cyclic and dynamic loadings are, at best, variable. Structures that are subject to wind loading are particularly difficult to analyse.
- *Stress analysis*: although there are well proven stress analysis techniques for the common structural shapes, these invariably involve simplification of the very difficult stress cases that can exist, even in a simple structural joint. Where several stresses are superimposed on a complex structure it is necessary to accept that some parts of it will be *indeterminate* and that accurate stress levels can only be found by empirical techniques such as photoelastic or strain gauge analysis.
- *Manufacturing variation*: variations occur in both the properties of materials, and the techniques used to fabricate them. The purpose of technical standards and codes of practice is to reduce the risks in these areas, but they can only ever be partially successful. Items such as castings and very complicated fabricated structures tend to be the worst – some risk of failure will always be present.

Factors of safety provide a further 'buffer' against this combination of technical risks, although often at the expense of economy. A straightforward land-based structure, where weight is not a prohibitive issue, will typically be specified with a safety factor of between seven and ten. The extra cost is seen as being offset by the future costs of safety and integrity problems that would result from using significantly lower values.

The Rainbow*: designing for manufacture*

It is common for the design of large steel structures to be influenced quite heavily by the practicalities of manufacturing. Fabricating structural steelwork is a low-technology activity which follows well-established practice. You can expect therefore that innovative design will be *constrained* by manufacturing considerations, not encouraged. The *Rainbow* has three areas in which design is affected: shape, cross-section and fabrication.

Shape   The original architects' sketch for the *Rainbow* indicated a true parabola and cutting this down by approximately half at the conceptual engineering stage retains the parabolic shape. The parabola is a well-known mathematical shape with the coordinate relationship $y^2 = 4ax$, where a is a constant. During manufacture the steel sheets (often termed 'flats') that will be fabricated to form the beam need to be bent to shape. This is done using a cold-rolling process; the flat is passed between several sets of rollers, under pressure, which bends the steel to a predetermined radius. It is possible, using an advanced rolling machine, to produce complex curve profiles. This is not a simple process, however, as it is necessary to compensate for the continual 'spring-back' of the flat caused by its elasticity. Practically, it is *much easier* – and cheaper – to roll the flat to an arc of constant radius. Figure 7.10 shows the actual shape of the beam. Note how the bottom section, approximately 6 metres long, is straight, rather than curved. The remainder of the beam comprises sections of three separate radii: 37.5 m, 9.0 m and 10.0 m as shown in the figure. This of course means that the final shape is not truly parabolic; however, if you plot all the coordinates and compare them with the theoretical $y^2 = 4ax$ profile, the maximum error is less than 0.2 metres and visually this difference is negligible.

Cross-section   The triangular section is built up from three separate flats joined by single-sided fillet welds. It is just feasible that the construction could be reduced to only two flats, but this would mean bending one of them into a V-shape to make up the triangular sides. Cracking at the bend would be a problem, as would rolling the V-shaped section to the correct radius. The thickness of the material and the section size increase towards the lower end of the beam, to compensate for higher stresses in this region. This raises the minor problem of joining metals of different thickness – the difference is small, however, and is not easily visible on the finished sculpture. A similar rolling technique is used for all the sections of the beam.

Fabrication   The *Rainbow* beam is fabricated in six separate pieces. This is purely for ease of manufacturing, handling and transport. The assembled sculpture is 18.5 metres high and could (just) be transported by road in one piece, but it is much easier in sections. The various radii are also a consideration in deciding how many

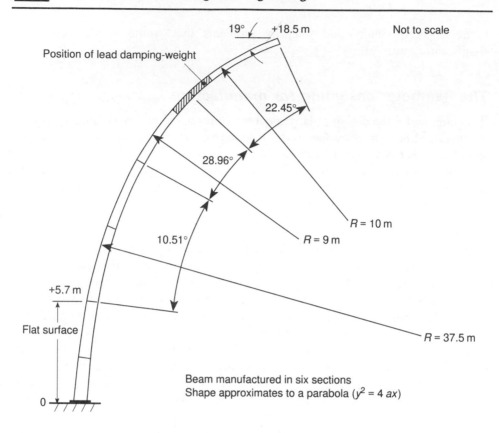

19° / +18.5 m   Not to scale

Position of lead damping-weight

22.45°

28.96°

$R = 10$ m

$R = 9$ m

10.51°

$R = 37.5$ m

+5.7 m

Flat surface

Beam manufactured in six sections
Shape approximates to a parabola ($y^2 = 4\,ax$)

0

Figure 7.10 Beam dimensions

separate pieces to use. The welding of the components follows common practice given in technical standards such as BS 5135 (ref. 2). For strength purposes, all structural welds are 'full penetration', in which the weld metal penetrates fully the parent metal. This gives a joint which is as strong as the parent metal. Note the weld preparation detail shown in Fig. 7.11; two of the welds that join adjacent sections are double-sided. This gives the strongest joint, but needs access to both sides. The final 'closing weld' is single-sided, because it can only be accessed from the outside.

7.4 A review

We have looked, in this case study, at some of the practicalities of working with an innovative design. But remember the question posed at the very beginning – where does all this innovation come from? Are we anywhere near finding the answer? The case of the *Rainbow* can at least provide some guidelines, some pointers. Think about, for instance, the *sources* of the innovation that appeared during the three stages of the design process. Most came from the conceptual and embodiment design phases, in surprisingly small steps. The conceptual idea of reducing the size of the

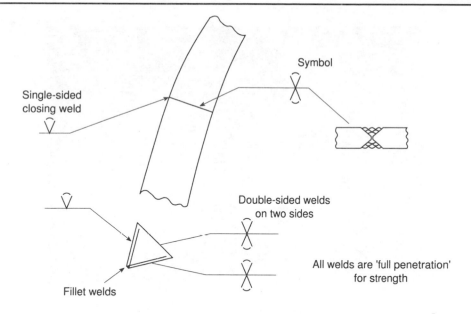

Figure 7.11 Beam construction – weld details

arc by half is a good example: Fig. 7.12 shows the final design. You can also see that some of the design process is not innovative at all – most of the detailed design features are well proven, it is just necessary to choose the correct ones to use for this innovatively shaped sculpture. So design innovation *is not* normally a long procession of clever, innovative steps; it is more often a few key, innovative steps and a lot of proven principles. Does this help us answer the question – the one about innovation? We did look at this, on a more generic level (when we weren't confused by the practicalities of engineering design), when looking at problem methodology, in Chapter 2. One of the conclusions was that innovation comes from *you*, not some 'other designer' (whoever that may be) – then we looked at the *Rainbow* and saw how innovation, in its nice small manageable pieces, is the bedrock of the design process. Here are the two key ideas again:

- Innovation is the bedrock of the design process; and
- Where does innovation come from? It comes from *you*.

7.5 Case study tasks

The objective of this case study is to help you gain an appreciation and understanding of the way innovation is handled as part of the engineering design process. The subsequent detailed design calculations are a necessary part of the process and are well covered by traditional textbooks. The following three sets of questions are designed to encourage you to think *forward* into the design process – design involves quite a lot of this, if the job is to be done well. You will need to understand the

Figure 7.12 The *Rainbow* sculpture – final design
(courtesy Avesta Ltd, Stockholm)

contents of the case study text and the basic messages about problem structure and methodology that were introduced in Chapter 2. Aim for individual answers to the questions rather than thinking of them as a full group exercise; you can discuss the points with others but the case is straightforward enough to warrant an individual approach. You can answer the questions in note form. There is no need for long-winded answers – it is the content that is important in this case, not the structure of your writing.

Attempt these questions

Set 1: about innovation

Q1:   Draw a schematic representation of how innovation *fits into* the structure of the design process. Explain where it has most effect.

Q2:   Make a list of all the ideas, activities and 'engineering' features mentioned in the case study that you would class as innovative. Include a one-line statement on each, explaining why you consider it innovative.

Set 2: about the design process itself

Q3:   When discussing problem-solving methodologies in Chapter 2 we decided that engineering design problems are often nested and/or iterative. Can you identify these properties in the design of the *Rainbow* sculpture? Explain, in note form, where they are.

Q4:   Good design involves anticipating how things might *fail*. List the possible failure mechanisms for the *Rainbow* sculpture. For each, identify a single design feature that could help reduce the risk of the failure.

Set 3: engineering considerations – stainless steel

Q5.   Why is stainless steel resistant to corrosion?

Q6·   What is the main problem caused by the welding of stainless steel?

References

7.1 BS 8100 (1986) *Lattice towers and masts*. British Standards Institution, London. This standard includes codes of practice for calculating loads.

7.2 BS 5135 (1984) *Specification for arc welding of carbon and carbon manganese steels*. British Standards Institution, London.

8 The *Inshallah* condenser – ISO 9000 application

Keywords

Principles of quality management – ISO 9001 – audits and certification – limitations. Specimen problem – tube failures – technical assessment versus QA assessment – application of ISO 9004-1 – the use of assessment checklists.

8.1 Objectives

This case study has a mainly procedural rather than technical basis. It looks in some detail at the way that the quality management standard ISO 9000 (ref. 1) is used as an assessment tool. To do this it is necessary to understand the structure and content of the standards and then be able to decide which of the various sections and clauses within it have most relevance to the job in hand. For this reason we will look at a product-specific assessment, rather than an overall 'company capability' assessment, which needs a slightly different emphasis. The case study will show you the advantages of using ISO 9000 as a way to help structure a design or product assessment.

Background: quality management

The ISO 9000 series of standards were first published in 1987 as an attempt to embody a set of comprehensive quality management concepts and guidance. They were designed with the principle of 'universal acceptance' in mind – and to be capable of being developed in a flexible way to suit the needs of many different businesses and industries. The series was developed in the form of a number of sections, termed 'models' – intended to fit in broadly with the way that industry is structured. Inevitably, this has resulted in a high level of *generalisation*. The current 'models' are as follows:

- ISO 9001 (1994) *Quality Systems – Specification for design, development, production, installation and servicing.*
- ISO 9002 (1994) *Quality Systems – Specification for production, installation and servicing.*
- ISO 9003 (1994) *Quality Systems – Specification for final inspection and test.*

In practice, you will find that these parts of the series are 'nested', i.e. 9001 contains all that is in 9002 and 9003 with some extra content. It is important to realise that these standards are not all different, but variations on a theme. The overall series is commonly referred to as 'ISO 9000' and is available from the British Standards Institution (BSI) – the address is given at the end of this book.

These standards have two key features. Firstly, they are all about *documentation*. This means that everything written in the standard refers to a specific document – the scope of documentation is, as you would expect, very wide. It is possible to go one step further and say that the direct requirements of ISO 9000 are *only* about documentation. This does not mean that they don't have an effect on the product or service produced by a company – merely that these are not controlled directly by what is mentioned in the standard. Secondly, ISO 9000 is about the effectiveness of a quality *management* system – it does not impinge directly on the design, or the usefulness, or the fitness for purpose of the product produced. It is a quality *management* standard, not a product conformity standard. It is, therefore, entirely possible for a manufacturer with a fully compliant ISO 9000 system installed and working, to make, and continue to make, a poor product. You may feel that this is a paradox but it is one that we have to work with.

Look at the contents list of ISO 9001, which is the most comprehensive and widely applied standard in the series. It is divided into four discrete sections: 0: 'Introduction', 1: 'Scope', 2: 'References' and 3: 'Definitions' are all short and serve mainly as an introduction. The main part is Section 4 which is divided into 20 clauses, some of which are in turn split into several subclauses. These 20 clauses form the core content of a good quality management system – you can think of them as being a list of *elements* of the quality management system – i.e. *quality system elements*. Figure 8.1 gives a list of these elements.

4.1	Management responsibility
4.2	Quality system
4.3	Contract review
4.4	Design control
4.5	Document and data control
4.6	Purchasing
4.7	Control of customer-supplied product
4.8	Product identification and traceability
4.9	Process control
4.10	Inspection and testing
4.11	Control of inspection measuring and test equipment
4.12	Inspection and test status
4.13	Control of non-conforming product
4.14	Corrective and preventive actions
4.15	Handling, storage, packing, preservation and delivery
4.16	Control of quality records
4.17	Internal quality audits
4.18	Training
4.19	Servicing
4.20	Statistical techniques

Figure 8.1 The 20 clauses of ISO 9001

Audits and certification

One of the quality system elements: 4.17 'Internal quality audits', is worth mentioning separately. A key property of a quality management system is the ability of the system to obtain feedback about the effectiveness of its own operation. To do this the standard specifies that regular scheduled *internal audits* are carried out by the company's own staff. An audit involves checking that all parts of the standard (i.e. the 20 clauses in Section 4) are being implemented. This requires that all the various documents are in place.

As an extension to this activity, the objective of most quality-conscious companies is to achieve *certification* of their quality management system. This is carried out by any of a large number of accredited certification bodies – they perform external audits on the content of the system and look at the way it is operating. Their scope is similar, albeit more formal, to that carried out by the company itself under Clause 4.17.

Limitations

One of the very real limitations of ISO 9001 is the fact that it is does not deal directly with design or product conformity. In its purest form, as published, its main use is to guide *system assessments* – which is how internal auditors and certification bodies use it. In practice, however, it can be used as an invaluable tool for analysing procedures and 'events' (notably those involving failures) on a product-specific basis. This is possible, not by changing the content of the standard clauses, but by adapting the way in which they are used as part of an assessment or investigation. It is an exercise in interpretation and targeting – both help to increase understanding of the wide-ranging application of ISO 9001, if it is used properly.

8.2 The problem: cracked condenser tubes

Background

Steam condensers, used widely in power and process applications, are frequently cooled by seawater. Seawater is highly corrosive and condenser tubing materials have a finite lifetime. The most commonly used material is copper-nickel alloy (Cu/Ni) but this can suffer from galvanic corrosion problems, particularly if the manufacturing process is not well controlled.

By far the best material to use is a high purity titanium (Ti) alloy – this is highly resistant to most types of corrosion experienced in a seawater environment, but it is also the most expensive.

Ti condenser tubes are manufactured from cold-rolled drawn strip, seam-welded using a shielded arc process (see Fig. 8.2). The condenser tube-plate is drilled and precision reamed (to a tolerance of ± 0.1 mm) to accept the tube which is then expanded, in the cold condition, using a portable rolling tool which fits inside the end of the tube and expands it radially outwards. The amount of expansion is of critical importance – if it is too great (i.e. the reamed hole is too big) the tube will be over-expanded and will work-harden during the expansion process. This will weaken the material, and the tube will also be left too thin. Conversely, if the amount of

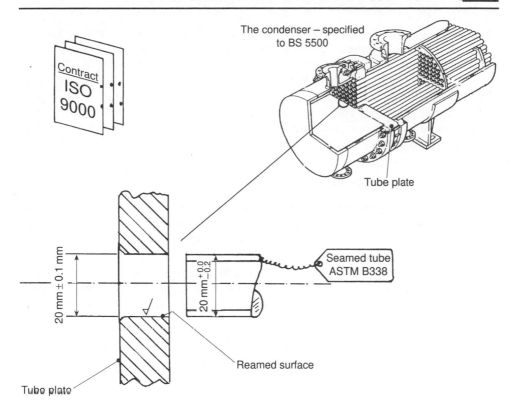

Figure 8.2 The condenser tubes

expansion is insufficient, there will not be enough 'nip' on the expanded joint and it will leak in service.

To monitor this key area a series of *expansion and flattening tests* are performed on the Ti tubes during their manufacture. Sample tube ends are expanded using a tapered drift pushed into the tube until the external diameter has expanded by 20 per cent. The expanded portion is then cut off and opened out for visual and microscopic examination to check for cracks in the material.

The titanium tube material is covered by the material specification: American Society for Testing of Materials (ASTM) B338 (ref. 2) which has been developed specifically for such applications. This is therefore not 'common stock' material. Typically, batches of tubes are traceable to manufacturing test certificates which provide a record of material analysis and mechanical test results. A common designation for these certificates is the European Standard BS EN 10204 (ref. 3). This defines the content of the certificates and specifies how the tests should be witnessed and the results verified.

The failure

Titanium condenser tubes to ASTM B338 were being used for the *Inshallah* chemical process plant. Unfortunately, half-way through the site assembly of the four large 20 000 tube condensers, the prevailing situation could probably not have been worse. Earlier material tests on the Ti tubes had (it was assumed) given acceptable results. A recently received set of laboratory test results however provided some cause for concern – the report indicated micro-cracks in the expanded and flattened samples. These cracks extended axially along the outside surface of the expanded tube portion and were well distributed around the circumference. The report explained that this was a classic work-hardening phenomenon. Also included in the accompanying laboratory report were the results of a chemical analysis test on one of the micro-cracked specimens.

Contract arrangements

These followed quite standard practice. The main condenser 'Design Contractor' had assumed responsibility for the detailed design of the condenser to BS 5500 (ref. 4) (a common pressure-vessel standard) and then subcontracted the actual manufacture of the components to several manufacturing companies. The Ti tubes were ordered from a specialist tube manufacturer, experienced with both seamless and welded tube, comprising ten batches each of 8000 identical tubes. By using ASTM titanium tubes, the Design Contractor felt he had complied fully with the end user's technical requirements as expressed in Clause X of the contract specification (see Fig. 8.3).

Clause (X): Condenser tubes
The main condenser tubes shall be manufactured of Titanium to ASTM B338. The quality of the material shall be to accepted international standards and the material shall have a system of identification and traceability. Tubes shall be located into their tubeplates using an approved method not requiring the use of seal weld. The completed condenser shall be subject to a gas leakage test to verify the integrity of the tube-to-tubeplate joints.

Clause (A): Quality assurance
The plant shall be designed, manufactured and tested to the requirements of the international quality assurance standard series ISO 9000 or its equivalent. The owner shall reserve the right to carry out audits and checks as deemed necessary to verify compliance with these arrangements.

Figure 8.3 Specification clauses – extract

The end-user's view

The end-user has become aware of the test result 'failures' and feels, understandably, somewhat uncomfortable that some or all of the 40 000 Ti condenser tubes already being assembled on site may be cracked. Secondly the end user's quality assurance department is worried that this whole contract is about to become a nightmare of supervision. As one operations manager put it:

> We decided after much discussion, and against the firm views of some of our own management, to specify these titanium alloy tubes in preference to the Cu/Ni ones that are less than half the price. And now look what's happened! We've got a half-built condenser full of cracks. It's going to be a real problem,

this *Inshallah* (literal meaning 'God Willing') plant. That's quality assurance for you.

The end-user's quality manager has raised the point that quality assurance is covered by Clause A, which is at the very beginning of the plant construction contract (Fig. 8.3). Together with his management board, he feels that the right thing to do is to call for an investigation – he has listed the following key points and objectives:

- Quantify the tube cracking problem. Metallurgy? Is it serious?
- Determine the extent of problem – how many tubes? Which ones?
- How did this happen (and can we make a financial claim against the Contractor for ignoring clauses A and X)?

8.3 Case study task

A careful reading of the case text should reveal that there are two 'nested' issues at work here. From the technical viewpoint there is clearly a need to check the Ti material specification compliance, the test procedures, and the test results, to find the *root cause* of the micro-cracking. To do this it is necessary to carry out a fairly comprehensive technical *design review*, including an assessment of the manufacturing activities. There is also a need to address the quality assurance aspects of the tube manufacture. This has to be seen to be carried out in some detail, given the apparent size and seriousness of the problem. It is likely that the company that supplied the Ti tubes has already been certified (or at least check-audited) to ISO 9001 as part of the vendor prequalification procedures for the project, so a more product-specific assessment is required. Your case study task is to decide the best way to structure the 'assessment' that has been requested by the plant's end-user. One way to do this is to follow these three steps:

Step 1   Having read the case, decide broadly how the assessment should be carried out, and what it is trying to prove (or disprove). Write a paragraph on *objectives* and then one covering *methodology*. Try to acknowledge briefly the views of the end-user, the contractor and the tube manufacturer – but don't drift too far away from the technical aspects.

Step 2   Draft a *checklist for the design review* (do not attempt the design review itself – you don't have enough information). Make this a checklist of specific technical questions that you would ask the various parties. It is important that questions be precise, rather than capable of being answered easily in a general sense. For example: look at this specimen checklist question on the subject of tube batches:

- 'Is the manufacture of the 80 000 Ti tubes divided into batches?'

This is far too general; the answer is almost guaranteed to be 'yes' and doesn't provide much enlightenment about any possible source of the cracking problem. A better version would be:

- 'How many Ti tubes make up a manufacturing batch that is capable of being separately identified – and how does the identification system work?'

Step 3 Compile a *quality assurance review checklist*. The 20 main clauses of ISO 9001 (see Fig. 8.1) are a good place to start but you will also need to use some of the subclauses of the standard. Again, this should be a checklist of questions which can be used during an actual assessment. It is also useful, where applicable, to note the specific documentation records, etc., that you would expect to see as part of the answer to a question. Use *only* questions that are relevant to the case study situation, as described, and refer to the clause numbers of ISO 9001 whenever possible. It is important to make the checklist questions complete and clear – a useful technique is to assume that someone else will have to use the checklist, rather than yourself.

References

8.1 ISO 9000: This is used as a generic name for the following series of quality management standards, published by the British Standards Institution, London.
ISO 9000–1 (1994) *Quality Management and quality assurance standards* – Part 1: *Guidelines for selection and use.*
ISO 9000–2 (1993) *Quality Management and quality assurance standards* – Part 2: *Generic guidelines for the application of ISO 9001, ISO 9002 and ISO 9003.*
ISO 9000–3 (1991) *Quality Management and quality assurance standards* – Part 3: *Guidelines for the application of ISO 9001 to the development, supply and maintenance of software.*
ISO 9002 (1994) *Quality systems – Model for quality assurance in production, installation and servicing.*
ISO 9003 (1994) *Quality systems – Model for quality assurance in final inspection and test.*
8.2 ASTM B338 (1994) *Standard specification for seamless and welded titanium and titanium alloy tubes for condensers and heat exchangers.* American Society for Testing of Materials, Philadelphia.
8.3 BS EN 10204 (1991) *Metallic products – types of inspection documents.* British Standards Institution, London.
8.4 BS 5500 (1994) *Specification for unfired fusion welded pressure vessels.* British Standards Institution, London.

9 Screwed fasteners – standardisation in design

Keywords

Standardisation and interchangeability – preferred numbers and preferred sizes – thread form toleranced dimensions – fundamental deviation – tolerance ranges – gauging and SPC. Design intent memorandum document – production versus quality control.

9.1 Objectives

It is difficult to find a production engineering process that does not involve some form of standardisation. In many areas standardisation is one of the core aspects of the manufacturing process – the lowest common denominator of mass production. Standardisation can exist at several different levels within a manufactured product. A typical motor vehicle, for example, incorporates over 600 separate standardisation systems. They range from, at the lower end of the scale, standard sizes of springs, screws, nuts and bolts, plastic clips and pipe fittings through standard bearings and pulleys to the final size, power rating (and price range) of the assembled vehicle itself. All of these involve standardisation systems.

The rationale of a standardisation system is simple – its purpose is to improve the overall economics of designing, manufacturing and selling the product by rationalising the range of sizes of components produced. This helps reduce waste and so increases the cost effectiveness of the overall manufacturing process. The basic elements of a standardisation system are also simple: they follow a well-established pattern, but they do involve a multidisciplinary approach utilising both mathematical and more empirically based engineering principles.

The objective of this case study is to show you how to develop a standardisation system, in this case for threaded fasteners, using the five main steps that are common to most systems of standardisation found in manufacture. More complex standardisation systems can be built by synthesis of these five main steps. The steps are:

- using preferred numbers/preferred sizes;
- deciding toleranced dimensions;
- setting tolerance values;
- incorporating manufacturing realities: measurement and inspection;
- defining design intent and design specifications.

9.2 Preferred sizes

It is a fair question to ask: 'preferred to what?'. Although standardisation has an empirical aspect to it, its basis is firmly rooted in mathematics, the starting point being series of *preferred numbers*. These are derived from geometric series, in which each term in a series is a uniform percentage larger than its predecessor. The five principal series (commonly termed the 'R' series) are shown in Table 9.1.

Table 9.1

Series	Basis	Ratio of terms (% increase)
R5	$5\sqrt{10}$	1.58 (58%)
R10	$10\sqrt{10}$	1.26 (26%)
R20	$20\sqrt{10}$	1.12 (12%)
R40	$40\sqrt{10}$	1.06 (6%)
R80	$80\sqrt{10}$	1.03 (3%)

Each of these series is capable of being expanded upwards or downwards by successively multiplying or dividing by a factor of 10. The numbers in each series are termed the preferred numbers (PNs) and these are commonly adopted as the basis for a range of sizes of components. Figure 9.1 shows, as an example, the development of the R5 and R10 series and the way that the PNs are normally displayed.

One problem with calculated PNs is that some of the terms (such as 3.15 in the R10 series) can be inconvenient to use, and imply a level of precision which is inappropriate or not required. Hence PN series are frequently subject to 'rounding off' to give a more rational grading. In practice, a maximum rounding of ± 1.5 per cent is sufficient to rationalise any of the five principal series. Clearly, the rounding off of PN series can produce several different results; for this reason it is necessary to consider also the practical aspects of using the PN size for a real engineering component. A PN series that has been rounded and rationalised is termed a series of

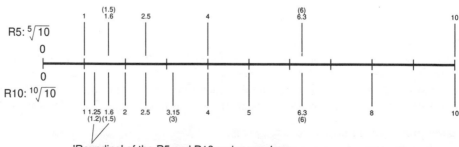

'Rounding' of the R5 and R10 series *numbers*
(shown in brackets) gives series of preferred *sizes*

preferred sizes. These sizes can then be used as a basis of the linear dimensions of a component.

Toleranced dimensions

In designing any engineering component it is necessary to decide which dimensions will be toleranced. This is predominantly an exercise in necessity – only those dimensions that *must* be tightly controlled, to preserve the functionality of the component, should be toleranced. Too many toleranced dimensions will increase significantly the manufacturing costs and may result in 'tolerance clash', where a dimension derived from other toleranced dimensions can have several contradictory values. Figure 9.2 shows good and bad principles of the choice of toleranced dimensions.

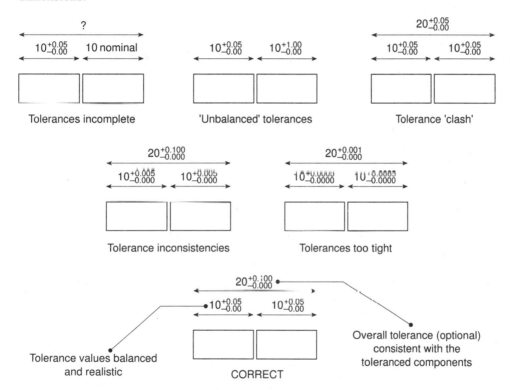

Figure 9.2 Toleranced dimensions – principles

Tolerance values

There is a well-established system of tolerancing adopted by British and International Standard Organisations and manufacturing industry. This system uses the two complementary elements of fundamental deviation and tolerance range to define fully the tolerance of a single component. It can be applied easily to components, such as screw threads, which join or mate together. The basic system is shown in Fig. 9.3.

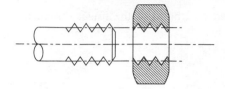

For screw threads, the tolerance layout shown applies to major, pitch and minor diameters (although the actual values will differ)

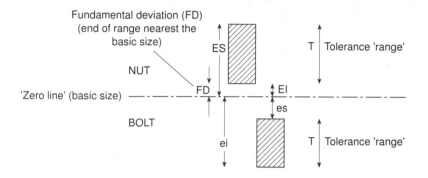

FD is designated by a letter code, eg: g, H
Tolerance range (T) is designated by a number code, eg 5, 6, 7

Commonly used symbols are:
EI – lower deviation (nut)
ES – upper deviation (nut)
ei – lower deviation (bolt)
es – upper deviation (bolt)

Figure 9.3 Method of tolerancing screw threads

Fundamental deviation   Fundamental deviation (FD) is the distance (or 'deviation') of the nearest 'end' of the tolerance band from the nominal or 'basic' size of a dimension.

Tolerance band   Tolerance band (or 'range') is the size of the tolerance band, i.e. the difference between the maximum and minimum acceptable size of a toleranced dimension. The size of the tolerance band, and the location of the FD, govern the system of limits and fits applied to mating parts.

Tolerance values have a key influence on the costs of a manufactured item so their choice must be seen in terms of economics as well as engineering practicality. Mass-produced items, by definition, are sold into markets which are both competitive and price sensitive, and over-tolerancing can affect the economics of a product range.

Measurement and inspection

Mass-produced components need to be measured and inspected at their point of manufacture to check for compliance with their design specification. These activities are often gathered together under the general discipline heading of *quality control.*

For standard engineering components a system of *sampling* is usually employed. A sample size is chosen, based on sound statistical principles, and subjected to an agreed inspection regime. Results are then extrapolated, again using statistical principles, to represent the results of the total inventory (the 'population') of the components. Such activities are described generally as *statistical process control* (SPC).

In order to perform inspection on an item, a system of *gauging* is generally required. Gauging is rarely used to measure all dimensions of a component, rather it concentrates on the toleranced dimensions. In practice it is rare for all toleranced dimensions to be gauged, as this would often involve multiple gauges, resulting in a very expensive and time-consuming activity. The solution is normally to select a number of key toleranced dimensions for gauging – which are indicative of the accuracy of the machining or other manufacturing activity that is being used. One of the most common methods of 'limit' gauging is the use of GO/NO-GO gauges. This is particularly useful for threaded components and those employing mating parts using a system of limits and fits. Two types of gauges are used. The GO gauge *must fit* over a component thread or dimension for the item to be acceptable, i.e. it checks the 'maximum metal condition'. The NO-GO gauge *must not fit* the dimension for the item to be acceptable, i.e. it checks the 'minimum metal condition'. An important principle utilised during GO/NO-GO gauging of screw threads is to restrict the gauging activity to the toleranced dimensions. Thread forms comprise a complex helical geometry and the checking of all dimensioned parameters is almost impossible under normal 'workshop' conditions. A simplified approach, relying on the assessment of three key toleranced dimensions, is well proven in industrial use. This requires the use of three separate gauging activities for internal (nut) threads and three for external (bolt) threads.

SPC relies on the underlying principle that the size distribution of mass-produced components follows particular (and predictable) patterns. There is no reason why screw threads should be an exception to this. Manufacturing variations such as tool wear, inaccurate mounting of steel stock in screwcutting/grinding machines and less predictable factors such as variations in material properties and temperature all contribute to the variability. Owing to the problems of direct measurement, an accurate assessment of *variability* of screw thread size is difficult. It is not normally possible, therefore, to obtain the full 'normal distribution' variability characteristic commonly used for components which have simpler linear dimensions. Limit gauging will only divide components into 'good' and 'defective' categories, so any statistical techniques that are used must be chosen to reflect this. Much depends on the *expected* number of defective items (i.e. that fail one or more of the GO/NO-GO tests) within the 'population' of components. This obviously varies with the 'quality' of the manufacture but for mass-produced turned or rolled screw threads a typical 'benchmark' figure is 2 per cent. For a pass/fail limit gauging technique, either a binomial $(q + p)^N$ or Poisson distribution based on exponential functions can be used. For large sample sizes containing a relatively small percentage of defectives, the binomial and Poisson distributions give similar results.

Practically, only a sample of each batch of items can be limit-gauged because of production time constraints. A common technique is to define an *acceptable* number of defectives to be found in each sample. If the actual number found is less than this agreed number, then the batch is defined as acceptable. If more are defective, the batch is laid aside for 100 per cent inspection. One essential characteristic of this approach is that the probability of a 'bad batch' being classed (mistakenly) as acceptable must be low. This type of SPC system can form a useful point of agreement between production and quality control departments in a manufacturing works.

9.3 The problem

There was no doubt that the Managing Director's presentation to the Board was professional: flowing dialogue, superb visual aids and good close logic. Almost glowing with pride, he presented details of the new order of 200 000 radial fan assemblies utilising new technology, shape and materials. Gradually, discussion moved round to the subject of cost. As well as the design cost of the new component shapes it would also be necessary to develop a new range of fasteners. The company's existing ranges of nuts and bolts, developed over several years' design and manufacture of larger axial fans, were too large and their square thread form just not suitable for these new precision assemblies. The Manufacturing Director confirmed that he was fully behind the 'in house' production of the fasteners; sizes could be carefully standardised and controlled and quality, as ever, would be the watchword. 'Quality', he had been heard to recite, was 'a journey rather than a destination'.

9.4 Case study tasks

As a design engineer it is your task (not the Manufacturing Director's) to design and specify the standard range of fasteners required for the radial fan assembly (Fig. 9.4). The *specification* element is an important part – the objective is to specify your design in a way that will be easily understood by 'shop floor' machine operators and inspectors. You will also need to document the methodology of your design decisions in a way that can be assimilated by other designers and managers; one way is by using a Design Intent Memorandum (DIM) document.

Methodology

You can structure your methodology using the broad guidelines provided in the case study text. These steps are shown in Fig. 9.5. The best first step is to assess the approximate range of fastener sizes (diameters) needed and match this to a preferred number series, rounding off to *preferred sizes*. Toleranced dimensions are set by first choosing fundamental deviation (try to match this to a PN series) and then calculating tolerance bands as indicated in the figure. You can use BS 2045 (ref. 1) and BS 3643 (ref. 2) to help you. For the purpose of the case study, only a single level of bolt/nut 'fit' is required. This means that you only have to derive and tabulate

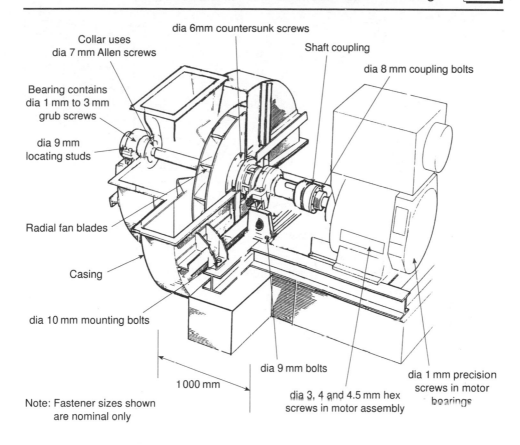

Collar uses
dia 7 mm Allen screws

dia 6mm countersunk screws

Shaft coupling

dia 8 mm coupling bolts

Bearing contains
dia 1 mm to 3 mm
grub screws

dia 9 mm
locating studs

Radial fan blades

Casing

dia 10 mm mounting bolts

1 000 mm

Note: Fastener sizes shown
are nominal only

dia 9 mm bolts

dia 3, 4 and 4.5 mm hex
screws in motor assembly

dia 1 mm precision
screws in motor
bearings

Figure 9.4 The radial fan assembly – fastener requirements

data for one external thread (bolt) and one internal thread (nut) tolerance grade, termed 6g and H6 respectively. Figures 9.6 and 9.7 provide basic information about sizes and gauging. Derive a simple system of GO/NO-GO gauge sizing and use statistical theory to define a sampling/inspection level for the finished range of fasteners.

One of the key aspects of this case study exercise is recording and communicating the design decisions that you have made – there is not one single 'correct' answer. The traditional way of recording design intentions is via a *design intent memorandum* (DIM). This contains design justification information, but without the fine detail. It is used to convey information to engineering managers and directors. At a more practical level, *design specification sheets* (DSSs) are used to hold the detailed manufacturing information. They contain sufficient information to enable the component details to be transferred to a shop-floor drawing for manufacture. The DSSs are also used by production planning departments to apportion manufacturing schedules and costs. Figure 9.8 shows a pro-forma DIM sheet. You should complete this as part of the case study exercise.

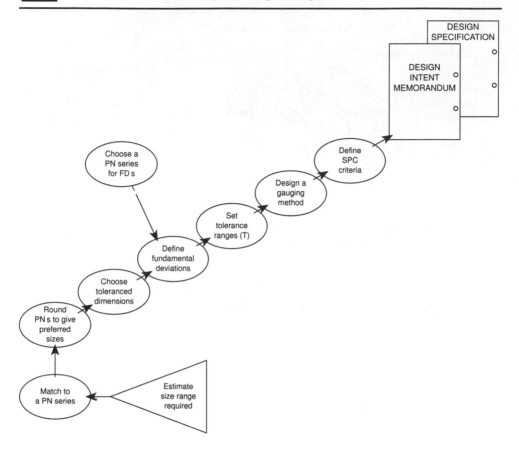

Figure 9.5 A methodology

Insight: production and quality control

In a small and active manufacturing company the disciplines of engineering design, production and quality control are closely interlinked. Nevertheless, the Production Manager here feels at the centre of the business:

'You must, after all, produce a fan assembly before you have one to sell'.

As Production Manager standardisation is her priority – given a free hand, she would make every component in the factory the same size:

'Too many sizes means unnecessary variations, problems and costs. And then there's the question of stock levels – currently the works only maintains three bar stock lines: 2 mm, 6 mm and 11 mm diameter, scrap levels are closely monitored and someone is always asking whether a few more fasteners can't be produced per kilogramme of rolled steel bar.'

She continued:

'Last year's high expenditure on machine tools made it one of the best (or worst) years for some time. These machines were chosen with some care to

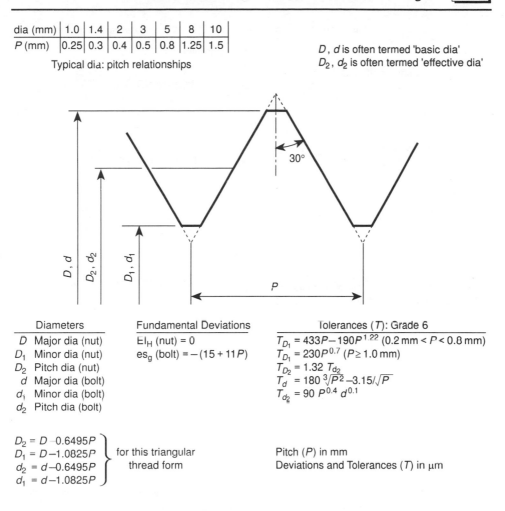

dia (mm)	1.0	1.4	2	3	5	8	10
P (mm)	0.25	0.3	0.4	0.5	0.8	1.25	1.5

Typical dia: pitch relationships

D, d is often termed 'basic dia'
D_2, d_2 is often termed 'effective dia'

30°

D, d D_2, d_2 D_1, d_1

P

Diameters	Fundamental Deviations	Tolerances (T): Grade 6

Diameters
D Major dia (nut)
D_1 Minor dia (nut)
D_2 Pitch dia (nut)
d Major dia (bolt)
d_1 Minor dia (bolt)
d_2 Pitch dia (bolt)

Fundamental Deviations
EI_H (nut) = 0
es_g (bolt) $= -(15 + 11P)$

Tolerances (T): Grade 6
$T_{D_1} = 433P - 190P^{1.22}$ (0.2 mm $< P <$ 0.8 mm)
$T_{D_1} = 230P^{0.7}$ ($P \geq 1.0$ mm)
$T_{D_2} = 1.32\ T_{d_2}$
$T_d = 180\ \sqrt[3]{P^2} - 3.15/\sqrt{P}$
$T_{d_2} = 90\ P^{0.4}\ d^{0.1}$

$D_2 = D - 0.6495P$
$D_1 = D - 1.0825P$
$d_2 = d - 0.6495P$
$d_1 = d - 1.0825P$
} for this triangular thread form

Pitch (P) in mm
Deviations and Tolerances (T) in μm

Figure 9.6 Basic thread-form data (metric)

produce the existing range of fastener sizes so any new size range will need to somehow "fit together" with the old one or serious questions will be asked, I can tell you.'

The Quality Control Manager proudly displayed his new matched set of precision screwthread ring and plug GO/NO-GO gauges:

'That will be half a minute per inspection per gauge, so let's see, for six gauges and a batch of 500 fasteners, that's just over 25 hours inspection time per batch – that'll need five of my inspectors, fully occupied. Inspection, that's what production engineering is all about.'

'I see', interrupted the Production Manager, 'interesting ideas; what do all these asterisks on the plan mean?'

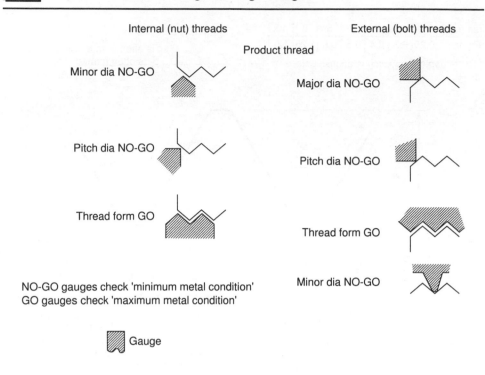

Figure 9.7 Thread gauging – principles

'Oh, they're to do with the quality levels, attributes, statistics, control charts and suchlike, all in accordance with BS 6000.'

'Fine, but it's my job to maintain utilisation on these new machines. Look, they have a standard production time of twenty seconds per fastener. That's just under three hours to manufacture a batch of 500 so, allowing for some margin you'll only have about two hours to inspect whatever it is that you have to inspect. Can't you do a sample, say 5 per cent?'

The Quality Control Manager (suddenly beset with the vision of only one of his five-person inspection team being usefully occupied):

'I'm not sure 5 per cent will be enough – just look at those tight tolerances: tools do wear, you know!'

Five per cent and 100 per cent inspection levels soon took on almost a life of their own. Two hours later, it was decided, jointly, to look at the design intent memorandum to see what *it* said.

References

9.1 BS 2045 (1982) *Preferred numbers*. British Standards Institution, London. This is an equivalent standard to ISO 3.

9.2 BS 3643 (1981) *ISO metric screw threads*. Part 1 – *Principles and basic data*; Part 2 – *Specification for selected limits of size*. British Standards Institution, London.

DESIGN INTENT MEMORANDUM

Subject: 1.

Fastener size range: 2.

Benchmark dimension: 3.

PN series chosen: 4.   5.

Maximum rounding error: 6.

Preferred size series chosen: 7.
(rounded)

8.

Toleranced dimensions: 9.

Fundamental deviation system used: 10.

Tolerance system used: 11.   Fit Class: 12.

Gauging system used: 13.

Inspection system used: 14.

DIM prepared by:	Relevant technical standards...............	See Design Specification sheet
Date	No. 001		No. 002

Figure 9.8 Proforma – design intent memorandum (DIM)

10 Fasteners and couplings – better design

Keywords

What is better design? – a methodology for design improvement (function, materials, manufacture, aesthetics). Threaded fasteners – failures – fatigue limit – thinking about improvements – a revised fastener design. Improving the design of a keyed coupling.

10.1 Objectives

You might hear a designer say something like this:

> Here is a new design of threaded fastener. I believe that it is better than the others because it has a more effective design. The design is more effective because. . .

Engineering designs are not static. Look at most engineering components and you will see that they change, over time, in shape and function. Some changes are large and sudden, others move step by step. The purpose of change is to improve design, to make the product more effective or, put simply, just plain *better*. But what is better design? Better design may be that which makes a product go faster, more slowly, makes it lighter, more reliable, harder, safer or cheaper – the list is very long and it would not be feasible to list them all. So how can we complete the statement above? The purpose of this case study is to show you the techniques of studying the mechanical design of a component so that you can improve it – make it more *effective*. The techniques are used in design studies in many industries; it is the principles and approach that are important.

What is effective design?

We are still trying to complete the unfinished statement. A design which is more effective:

- does its job better; and
- gives better value for money, over time.

These two statements seem useful, but they are limited because they are lowest common denominators. In this case study we will look critically at various principles and practices that are used within the design of a component. This is a basic, commonplace design activity, often called a 'design study' or 'design review'. These activities happen when an existing product needs to be improved. Do not confuse this with the innovative or brainstorming phase very early in a product's conception when detailed mechanical design takes second place to general concepts (if you want to see how this phase works, read Chapter 7). Here, we are concerned with changes in the *detailed* mechanical design of a component.

Structure

You need a structure for two reasons. First, change is difficult and often controversial – there will be differences of opinion about whether your design changes are desirable or not. A structured approach can help keep things in focus. Second, mechanical design is complicated, it has many facets and so it makes good sense to sort these into a logical order. Figure 10.1 shows the structural parts of a mechanical design review. It breaks into four questions:

- What does the component do? (function)
- What is it made of? (materials)
- How is it made? (manufacture)
- What does it look like? (aesthetics)

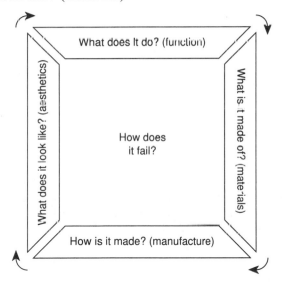

Figure 10.1 The parts of a mechanical design review

Note how the four legs of the structure form a loop, joining together as a continuous process of review and (hopefully) improvement. Having set the structure, we now need to see how it is used in practice. We will use, as an example, simple threaded fasteners. We will review the design and improve it. Somewhere between these two activities we will need to tease out the key design issues.

10.2 The problem: fastener failures

Threaded fasteners are used for many applications so although their function is simple – to hold components together – there is a wide variety of designs and sizes.

There are specific designs for applications where the fastener needs to be very strong, such as for machine applications. Fastener design changes slowly, most development being made for specialist fasteners where changes are generally to do with thread form or materials of construction. One key factor is that fasteners are produced in large numbers, so developments that are made must be compatible with the techniques of mass production.

The big problem with threaded fasteners is that they *break*. This normally happens when a tightened fastener has been in position for some time and is particularly common when the assembly is subject to cyclic loadings, such as vibration. A secondary problem is the potential for fasteners to work loose, although various locking devices are available to prevent this, so it is not a serious design problem. Breakage is the main failure mode; hence the main objective of a design improvement programme is to develop a fastener design which will have a lower chance of failure under typical operating conditions. To do this it is necessary to look at both parts of the fastener: the nut and the bolt.

10.3 Technical design

It is best to consider technical design in relation to a simple methodology. This is not essential for a simple component such as a fastener but it becomes important for larger or more complex equipment – so it is good practice to become familiar with using a structured approach. The methodology is shown in use in Fig. 10.2. We can look at each leg in turn.

Component function

Note how the component function has been broken down into a few basic descriptions of what the fastener *does*: the next step is to draw some engineering inferences from these descriptions, as shown in Table 10.1.

An important function point not shown in Fig. 10.2 is the way that the nut and bolt fit together. Fastener design standards specify that eight threads are in engagement (you can check this using any standard nut and bolt) and that the thread fit should allow a standard set of fit classes (i.e. the g/H designations used in Chapters 6 and 9).

Materials

The steel used to make fasteners can be divided broadly into low and high tensile grades. In use it is not only the tensile strength that it important; the fastener also relies on:

- compressive and shear strength (of the nut);
- toughness, to resist impact loadings;
- hardness to resist surface indentation;
- corrosion resistance so that the nut or bolt body and their threaded fit are not weakened by rusting;
- fatigue strength.

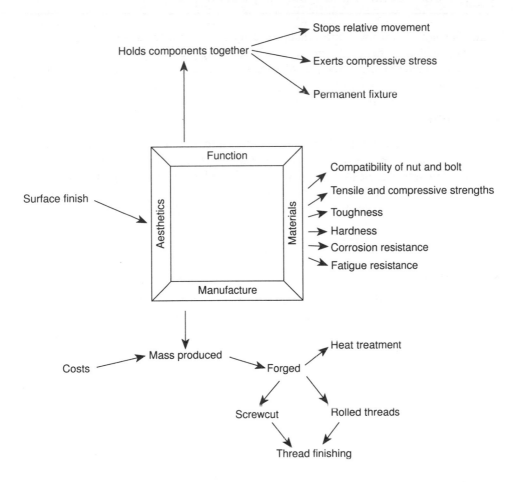

Figure 10.2 Methodology for threaded fasteners

Table 10.1

Function	Inference
It holds assembled components together, so . . .	It exerts a mean compressive stress – hence the bolt is in tension and the nut is in *compression*.
It stops relative movement, so . . .	There will be an imposed bending stress if the fastener is not close-fitting.
It remains in place for a long time, so . . .	It will be subject to fatigue loadings.
There are two separate components, nut and bolt, so . . .	Dimensional 'fitting' tolerances will be important.

Some (but not all) of these are related directly to the resistance of the material to plane tensile stress – hence the nominal tensile strength and modulus of elasticity of the fastener material are only partial measures of how suitable it will be for a particular use.

Manufacture

Mass-produced fasteners smaller than about 5 mm in diameter are produced by *rolling* the thread form, using matching, profiled rolls. Larger fasteners are made by *upset forging* the bolt head from a steel blank and then producing the thread by lathework cutting. Final thread profile is achieved by an accurate screw-cutting process called *chasing*. Fasteners are designed for mass-production so they have to be made to close tolerances to give interchangeability. Owing to the large number of fasteners made, they must be able to be made *economically*.

Aesthetics

Aesthetics is not a major consideration for industrial fasteners, but there are several related considerations which have a practical effect. The nut and bolt hexagon heads must fit standard tools and it is necessary to have a wide range of preferred sizes to meet commercial applications. Surface finish is also important for fastener fitting and accuracy, as well as cosmetic appearance.

How fasteners fail

You can learn a lot about how to improve a component's mechanical design by studying how it *fails*. Failures in threaded fasteners occur mainly in the bolt – in one of the three common places (shown in Fig. 10.3):

- at the point where the unthreaded part of the bolt shank meets the hexagon head: this is called 'underhead fillet failure';
- at the point where the threaded part of the bolt enters the nut, i.e. at the end of engagement;
- at the end-point of the bolt thread, i.e. the *thread start* nearest the head-end of the bolt.

Now for a little clear thinking – what do these three failures have in common? The first check should be to investigate whether these failures are caused by excessive plane stress. If this was true there is not a workable explanation of why the bolt breaks specifically at the end of engagement; it could break anywhere along the threaded portion of the shank where the root diameter of the thread produces an increased plane stress. Other clues also lie with the location of the underhead fillet failures and the thread start failure – these are areas of stress concentration. The thread start failure occurs at a *notch* in the material; this fits in with the common engineering principle of notch failure (known more formally as *notched-specimen failure*). Together, the observations point to a clear engineering conclusion:

- threaded bolts fail by a *fatigue* mechanism, not because the mean plane stress level is too high.

The mechanics of this are shown at the bottom of Fig. 10.3. The mean tensile stress (σ_m) is accompanied by a fluctuating stress of amplitude $2\sigma_a$, often caused by

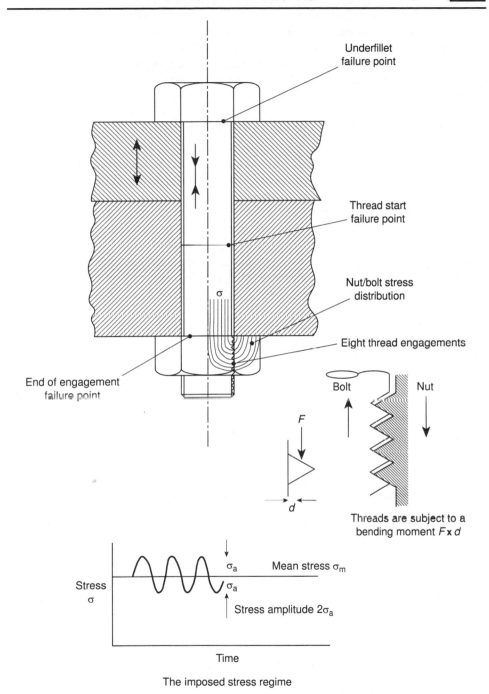

Underfillet
failure point

Thread start
failure point

Nut/bolt stress
distribution

Eight thread engagements

End of engagement
failure point

Bolt   Nut

σ

F

d

Threads are subject to a
bending moment $F \times d$

Stress
σ

σ_a Mean stress σ_m
σ_a

Stress amplitude $2\sigma_a$

Time

The imposed stress regime

Figure 10.3 Fastener stresses and failure points

machine vibration. Look at the other design inferences in this figure; the lines of stress concentration around the nut/bolt thread interface show why stresses are higher in the region around the end of engagement and the first nut thread. A quick analysis of the thread contact should show you how the threads are subject to a bending force, and that the highest stress will occur at the thread root. These are useful pointers towards possible design improvements. The main issue though is fatigue strength, so the next step is to look at this in more engineering detail.

10.4 Thinking about design improvements

Now that we have identified fatigue strength as the most important design issue, the technical picture starts to become clearer. A good technique is to look for a match between actual observations and well-accepted engineering theory and practice. There is a link here to the type of 'closed problem' described in Chapter 2 – the objective is to *open up* the problem, to reveal its technical facets. This will help progress towards the desirable design improvements. Figure 10.4 shows how this methodology has been applied to the issue of fatigue strength. Note the general progression using the three linked steps: fact, development and conclusion.

Fact	Development	Conclusion
The *notch sensitivity* of a material increases as tensile strength increases.	Fatigue limit is not increased significantly by using a stronger material.	Using a high-tensile bolt material will not necessarily prevent breakage.
The fatigue limit of a notched specimen decreases with specimen size.	This results in a 'size effect' for bolts as they behave as notched specimens.	Increasing a bolt's diameter will not always prolong its life.
Many bolts fail at mean stresses significantly lower than plane yield stress.	Mean stress (σ_m) is a secondary issue in fastener design.	Reducing the σ_m will not prevent breakage.

Figure 10.4 Thinking about design improvement – technical analysis

Fatigue limit

A key point in engineering design is the concept of a *fatigue limit*. This is the maximum stress amplitude ($2\sigma_a$) range that a material can withstand over an (in theory) infinite number of cycles without failure. The 'facts' shown in Fig. 10.4 are generally accepted – you can find them in engineering handbooks and references – and show how conventional engineering wisdom is *brought to the problem*. Note how three specific conclusions about design improvements have been reached by developing these factual ideas.

Thread interaction

The fact that bolt failures can occur at the end of thread engagement gives a clue that the thread itself is causing a stress concentration. A little engineering design *thinking* will show you why this happens. Look at this in two ways: from the viewpoint of load *distribution* on the threads, and then at the stress acting on an individual thread.

Thread load distribution   Each thread in a nut/bolt engagement does not carry the same load. The first thread carries up to 30 per cent of the load and the percentage reduces along the engagement – the last four threads carry only about 10 per cent between them. This is due to the elasticity of the material and the influence of localised loadings between the contacting surfaces. Figure 10.5 shows the distribution for a standard eight-thread engagement. This is inefficient design: the first few threads are taking most of the stress, the others acting almost only as a 'guide'. It is clear how the increased stress concentration resulting from the uneven load-sharing will increase the stress near the end of engagement and hence give this area a lower fatigue limit than the rest of the bolt shank. Better load sharing between the threads would be a design improvement.

Thread stress   In use, when the nut is tightened, there is a shear force acting between the engaged threads of the nut and bolt (look back at Fig. 10.3). From here, it is a short step to the assumption that each thread is loaded as a cantilever, restrained at its root by the root diameter of the nut or bolt material, and then subject to a bending force imposed by its mating thread. We can assume that contact occurs at the 'effective diameter' (near the mid-point of the thread height), giving a bending *moment*. Now move one step further – the effect of such a bending moment will cause maximum stress to occur at the *root* of the thread form.

Is this a useful conclusion? We could conceive that an increased thread root radius would decrease the stress level and so make the threads stronger. This is true, in theory. The problem comes when you have to consider the practicalities of the idea

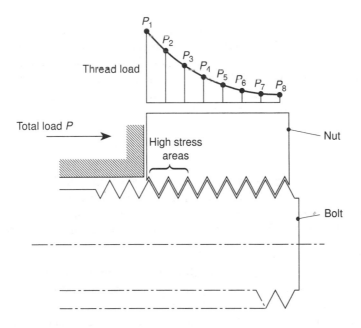

Figure 10.5 Thread load distribution – inefficient design

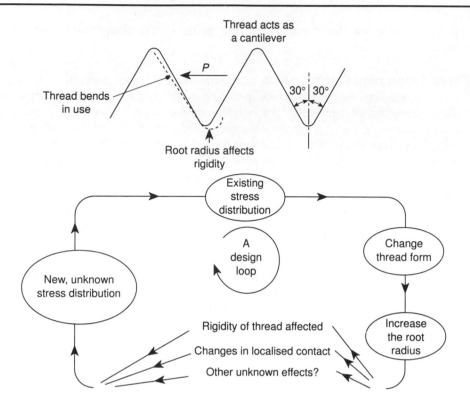

Figure 10.6 Try to avoid design *loops*

— a larger root radius would also increase the amount of localised contact between adjacent mating threads. This *could* change the entire stress regime, so it is not clear whether this would be an improvement or not. The situation is summarised in Fig. 10.6 — note the potential for a design 'loop' here. Loops are a common danger of design improvement attempts; it is best (but not always possible) to avoid them or at least recognise when you are dealing with them.

Materials

So far we have reached the conclusion, from consideration of the fatigue limit, that using a higher-strength nut or bolt material is probably *not* a design improvement. Consider the alternative scenario, the use of a softer material — it is difficult to see how this could be of benefit if used for the bolt, but what about the nut? Now develop this idea:

- Contact between thread faces is localised because of machining inaccuracies.
- The nut and bolt are made separately so there will be variation in dimensions. This increases the chance of localised contact.
- Localised contact produces high stresses and unpredictable stress concentrations.

- A softer nut would enable the nut threads to deform, giving a greater contact area. This would reduce stress and help to stop the nut coming loose.

Look how these ideas have developed – the technical correlations between the four statements are not perfect but there is a broad technical progression. Note also how the problem of fatigue limit is being 'opened up'. This is a good example of how to treat a *closed problem* – a concept introduced in Chapter 2 of this book.

Notches

As a final effort in 'improvement thinking', why not try to eliminate the half-thread? It is a known cause of failure because of the way that it reduces the fatigue limit on the bolt. If we retain the parallel shank, there are only two ways of eliminating this half-thread:

- continue the thread all the way up the shank to the bolt head;
- recess the unthreaded portion of the shank.

Ending the thread at the bolt head will make things worse, as this is already an area of stress concentration failure. The second idea, recessing the unthreaded portion of the shank so the thread does not end in a notch, is a better solution.

To summarise the methodology so far: we have assessed the design function of threaded fasteners, looked forward to how they fail, and then teased out the technical issues using some clear technical thinking. We have homed in on fatigue strength as the most important technical issue and opened up the problem, looking for design areas that need to be improved. The objective now is to synthesise these together into workable engineering recommendations. These need to be practical recommendations: it is of little use presenting the theoretical basis without the practical implications that go with it. The easiest way to do this is to first represent the situation, on a chart or drawing, listing the various improvement points as they relate to the four main 'legs' used in Figs 10.1 and 10.2. There is no absolute reason why you have to do it like this – but it can help by making things clear. Note that this is a *procedural* step; it does not express anything new.

Now revise the design. *Improve* it. Take the points from your list and turn them into engineering reality, using sketches, not text. Narrative is not expressive enough. Annotate the new design features in the sketches as you draw them until you have something resembling the layout and detail of Fig. 10.7. It is acceptable for the ideas to remain flexible at this stage; note some of the 'offshoot ideas' in the figure. Separate sketches, as shown, can contain more detail; the idea is to assemble a collection of workable design improvements which can then be added together to produce a revised design. A typical improved design is shown in Fig. 10.7. There are many possibilities, as long as the result is consistent with the technical conclusions that we have found.

10.5 The improved design

The features of the new fastener design are as follows:

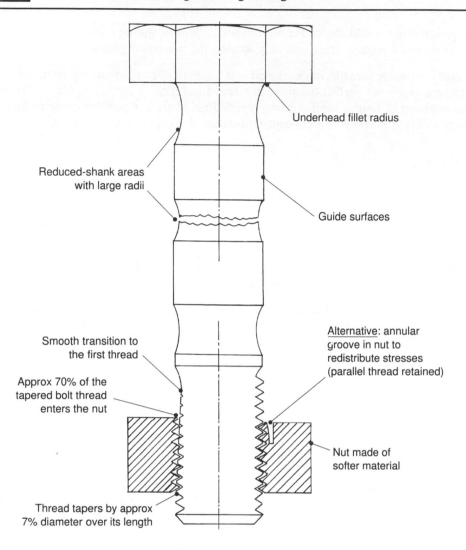

Figure 10.7 The improved design

Tapered thread

A small thread taper angle of 6–7° gives a more uniform stress distribution between the threads. The threads nearer the head-end are shorter and so have a lower bending moment on them.

Recessed bolt shank

The shank is recessed, eliminating the half-thread 'notch'. Radii are incorporated to reduce any resulting stress concentration around the changes of section. Note how the parallel sections of shank are retained to act as guides for the fitting of the bolt into the assembly. This has the effect of keeping mean stress levels down – and a clearance all the way along the shank would subject the bolt to an additional bending stress, which is undesirable. With the removal of the notches, the bolt will behave more like a plane specimen rather than a 'notched specimen'.

Nut material

The nut and bolt are made from dissimilar materials, the nut being softer to increase the contact area and reduce stress concentrations.

Nut shape

This is one of the more controversial design changes. The nut has an annular groove feature at its inner end, which serves to reduce the stress concentration around the first two threads. This works with the slight thread taper to reduce overall stress in this area. It is likely that this more complex nut shape will make manufacture a little more difficult and increase the unit cost.

Other issues were considered for the new fastener design, but have not resulted in design changes. It is worth noting these, and a brief justification is as follows:

- *Shank maximum diameter*. This remains the same. There seems little advantage to be gained by increasing the shank diameter to reduce the mean stress as this will not increase the fatigue limit much.
- *Dimensional accuracy*. This is unchanged. There is no evidence that increasing dimensional accuracy will increase the fatigue limit. The same is true of surface finish – although it is well known that a smooth surface finish increases the fatigue limit, this has already been incorporated into commercially available fasteners. If the surface finish was causing fatigue failures, then the location of the bolt shank failures would be more random than as described. A further point is that improvements in dimensional accuracy and surface finish are expensive because better tooling and longer manufacturing times are needed.
- *Prestressing*. The action of prestressing a bolt, by tightening it sufficiently to produce plastic deformation, is a well-known principle. It is claimed to increase the fatigue limit by changing the way in which the material reacts to fatigue (because it has been stressed above its elastic limit). This is a metallurgical issue rather than a design point *and* it does not fit all of the different sets of assumptions that can be used to describe material failures. From a pure fastener design viewpoint it is best avoided.
- *Manufacturing methods*. It is not advisable to get *too* involved in manufacturing methods at this level of design review and improvement. How an 'improved' component will be made is a key issue, but its effects are mainly on the economics of producing the component. This will divert attention from the whole idea of design improvements. The best results come from making active attempts to improve mechanical design, discussing and revising as necessary, but leaving serious consideration of manufacturing costs until later. If you spend too much time worrying about manufacturing costs you may end up making no design improvements at all.

10.6 Case study task

The design review of the threaded fasteners shows how to analyse logically the design of a simple component and *develop* design improvements. The objective of the case study task is to apply these techniques to a different engineering component – a simple keyed drive. This is a similar type of problem to that of the fasteners; although it is a moving, rather than static, component it shares many of the same engineering principles and some similar design features.

Your task is to produce an improved design for the keyed drive coupling shown in Fig. 10.8. This is a rotating shaft fitted to a heavy-duty overhead crane. The shaft transmits drive from an electric motor to a reduction gearbox which moves the crane hoist assembly along its girder 'bridge'. Overhead cranes carry heavy weights and have to be designed to resist vibration and shock loads. Basic design data are shown in the figure.

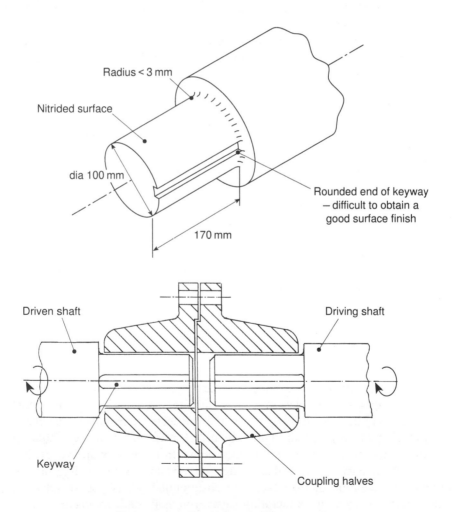

Figure 10.8 The keyed drive coupling

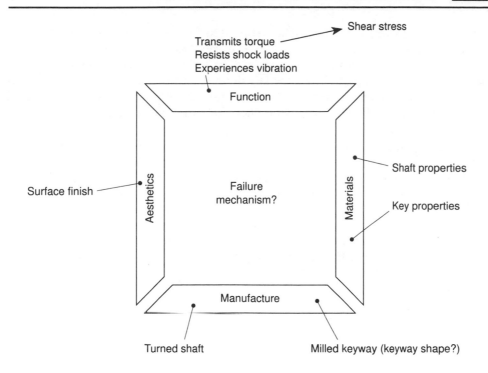

Figure 10.9 Your starting point

The following commentary provides outline information on the design of keyed couplings and how they fail. Use this as background information to help you identify desirable design improvements. You should specify your improved design using dimensional sketches, with important design features annotated. Detailed engineering drawings or lengthy narrative are not required – it is the *ideas* that are important. Try to follow the methodology used for the threaded fasteners described in the case study text – start with Fig. 10.9, completing it in a similar way to Fig. 10.2.

Keyed couplings – function

A keyed coupling is a common method used to transmit rotation between two shafts. Each shaft has a cast boss which fits over it using a medium clearance fit. The two boss halves are located accurately together by threaded fasteners. Drive is transmitted by a tight-fitting key, extending along the shaft, located in matching keyway slots in the shaft and boss. Shaft couplings have to be designed to withstand shock loads and vibration.

Design points

Coupling shafts are usually stepped so that larger diameter bearings can be used on the main body of the shaft. The slenderness ratio of the shaft is small ($^{L}/_{D} < 4$) which makes it stiff. The factor of safety of the shaft (approximately 4) is chosen to resist predominantly shear stress due to torsion but also the effect of shock loads when the shafts start or stop suddenly. The shaft is made of stainless steel with a limiting shear stress of 75 MN/m². A stronger material is used for the key, giving a shear stress limit of 90 MN/m². The most common type of key has a square section and is sized to be a close press fit in the shaft keyway. The fit in the matching boss keyway is loose, so the boss can be slid over the key during assembly. The shaft keyway is cut using a slot-mill cutter mounted in a vertical milling machine so it can be traversed along the shaft length. The shaft is produced by turning on a lathe.

How they fail

There is a common failure mode for keyed couplings. The shaft breaks radially at the point where the end of the keyway meets the shaft step. A typical fracture surface will show a smooth area around the outside of the shaft section, extending to a depth of about $d/3$, whilst the centre section has a rougher 'cauliflower-like' appearance. The coupling boss and the key itself rarely fail, even if the coupling bolts have come loose in service causing the assembly to vibrate.

11 *Piranha* – technology transfer and project structure

Keywords

Technology transfer projects – space vehicle construction – the problem of complexity – project organisation and structure – technical interfaces. BS 7000 *Managing product design* – quality management (ISO 9000) – methodology – defining quality system elements.

11.1 Objectives

This case study is about the problems of large design and manufacturing projects that are undertaken on an international basis and which incorporate *technology transfer*. Increasingly, complex machines in the fields of aerospace and space technology are made in this way. Over the past ten years the scenario whereby individual parts of such projects are designed and manufactured in different countries has become more common. Improved political links between mature industrialised and developing countries, with the consequent reduction of some technical and commercial barriers, have helped this to happen. The most common structure involves a major technology *licensor*, who owns the generic technology, acting in conjunction with a number of manufacturing *licensees* (who may also be involved in design and act, in turn, as sublicensors) who make the component parts. The structure operates on a *consortium* basis, rather than a more traditional buy/supply relationship. With such highly structured arrangements, effective project management becomes a necessity.

In concept this case study is about *organisation and structure*, considered within a system (i.e. a manufacturing project) that is predominantly technical. In detail it is about careful, precise definition of roles (who does what) at the project level but, once again, within a specific system of very real technical constraints. Two important concepts are introduced: *design* management and *quality* management. The concept of design management is central to technology-transfer projects – it is of little use manufacturing complex high technology components that will not fit together. Quality management has also made recent advancements; standards such as the ISO 9000 series and BS 7000 (ref. 1) are gaining international acceptance as a basis for a good quality management system which will work in many design and manufacturing situations.

This case study looks at the technology-transfer design and manufacturing arrangements for a rocket-launched space vehicle. It involves several aspects, from dividing the overall project into phases and convenient 'boxes of responsibility', to the identification and clear definition of technical interfaces. None of these aspects are particularly easy – they tend to be nested together, overlapping, and in some cases, multidisciplinary. You will need, therefore, to study and understand fully the basic concepts introduced in the case text before attempting the later case study tasks and exercises. Although the case discusses specifically space vehicle manufacture, the concepts and principles are general ones and have wide-ranging application to many complex technology transfer projects.

11.2 Space vehicle construction

A space vehicle, although comprising advanced technology, is essentially built the same way as any other vehicle or machine. It consists of an assembly of component parts, making up a structure, and a set of mechanical, electrical, hydraulic and pneumatic systems. A supporting set of 'information' systems provide operating support for control and process functions. Figure 11.1 shows the *Piranha* rocket vehicle and its major installed systems. The component parts and systems of the vehicle can be broken down into the following six groups.

The airframe – primary

The primary 'airframe' is the main structural envelope of the vehicle consisting basically of interlocking thin-walled tubular members strengthened by internal circumferential and axial ribs. Additional structural reinforcement is located in areas of high stress, such as around the rocket motor engine mounts. The materials used are specialised high strength and heat-resistant alloys. These alloys have been developed specifically for space vehicle use and so are not covered by the normal 'international' ranges of published technical standards. Detailed material properties and precise chemical analysis are considered 'classified information'. Once the material properties are known, however, the design of the main structural parts is actually quite simple and standard stress analysis techniques can be used to determine material section thickness, factors of safety, etc.

The airframe – secondary

These are the non-structural external parts of the vehicle's hull and comprise access hatches and non-pressurised components such as external control surfaces, flaps, and stabilisers. Although they play no part in supporting the weight of the vehicle structure, they are still subject to aerodynamic loads when the vehicle is in flight within the earth's atmosphere. As for the primary airframe components, special 'bespoke' materials are used.

The rocket motors and fuel system

Two separate rocket stages are used to propel *Piranha* into orbit. Each stage has two identical rocket motors burning liquid fuel, oxygen and hydrogen, which are fed by a centrifugal compressor into the combustion chambers where they are ignited. The

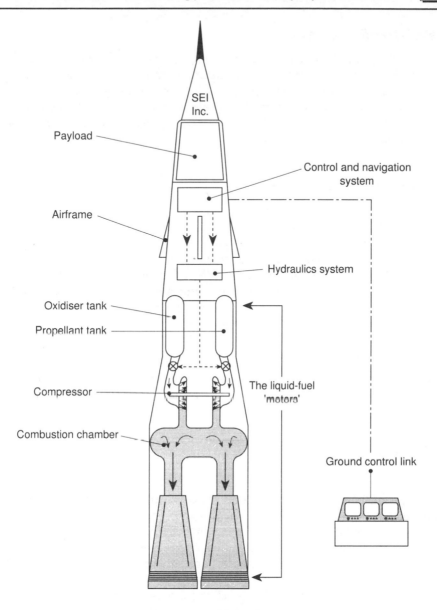

Figure 11.1 The *Piranha* rocket vehicle

pumps work on the turbocharger principle, using the kinetic and pressure energy of the expanding combustion products to drive the compressor via a turbine. Automatic valves control the flow of the propellants, enabling the motors to operate at various degrees of power. Although the *principle* of a rocket motor is straightforward, in reality it is a complex assembly of approximately 15 000 individual components. Many of these are small but important items such as seals, pins and locking devices. Accurate manufacture and assembly to fine tolerances is essential. Reliability of the motor is paramount, as is its physical integrity, so extensive laboratory and operational tests are required.

The hydraulic systems

The hydraulic systems are the nearest to a 'proprietary unit' that can be found on a space vehicle. Several discrete systems are installed to operate stabiliser surfaces, docking mechanisms and various actuators. These are designed on a 'modular basis' so many of the subsystems are interchangeable. The various system components are generally not bespoke designs – they are proprietary mass-produced items, made to well-established technical standards.

The computer hardware

The key aspect of computer hardware design is compatibility. The hardware is *all* of bespoke design and is a core part of the main licensor's technology base. Standard computers will not do the job. The bespoke computers contain about 95 per cent 'standard' electronic components but the remaining 5 per cent are of special design, and once again can be considered 'classified'. A further essential requirement is the need for *absolute compatibility* between the hardware and the software, to avoid any mismatches, however small, which could cause functional problems.

The computer software

Space vehicle software is completely bespoke and, like the hardware, forms a core part of the main licensor's technology package. The two main criteria of the software design are absolute compatibility between systems, and a successful programme of long-term testing to demonstrate the absence of common-mode failures due to contradictory programme entries. As part of the test programme, the 'as-installed' hardware and software need to be tested together.

The designs of the six main systems must be considered together within the overall concept of the space vehicle design. The design process for the main airframe and rocket motors is iterative, in that changes to airframe size, shape or weight invariably have an effect on the engine sizes required. This, in turn, requires changes to the design criteria of the fuel system. From the control aspect, the rocket motor control system must be considered almost 'as one' with the function of the hardware and software, to ensure full compatibility *is* achieved. As a general principle there is less design iteration necessary in this area. Final design solutions are normally arrived at reasonably quickly but the timescale is then extended by the need for combined functional tests. *Functionality* is the key point. There are many more statements that could be made regarding the interlinked and iterative nature of the design process for such a complex engineering product. There are also many more systems and subsystems that could be added to the picture if we increased the depth of our analysis. The main message is that the overall picture (the 'true' picture) is one of almost infinite complexity. Hence any analysis of it, or any attempt to manage it, must by definition be imperfect. This means that a framework to work to, a *conceptual structure*, is necessary – it is an essential, rather than optional, part of understanding technically what is going on in the design and manufacturing process.

11.3 Design and manufacture: contracts

The following is an outline description of the role of *contracts* in the design and manufacture of *Piranha*.

From the public's perception, the *Piranha* space programme is the brainchild of Space Exploration Industries (SEI) Inc., who design and manufacture the rocket vehicle and its exploration modules and payload communication satellites. In reality, this is partly true; SEI are the overall technology licensors for the completed vehicle; they consider it 'their' rocket vehicle, having spent ten years and almost one billion dollars on its development. Legally also, the scope of the SEI licence covers the complete assembled rocket vehicle. From a technical viewpoint, the situation is a little more complex. The computer hardware and software systems are fully designed and engineered by SEI in their own development laboratory and assembly factories. They have no 'partners' in this enterprise, and buy electronic components on a purely 'supply only' basis. Development of the 5 per cent bespoke electronic components, and assembly of the others, is carried out in the SEI factory. The design and manufacture of the rocket motors is subcontracted entirely to a specialist organisation who act as technology licensor for the engine design but subcontract the manufacture of all the auxiliaries such as the fuel system (which has strong design links with the computer hardware/software system) to other companies.

The airframe design criteria are set by SEI and issued as a functional specification *with* detailed classified material information, to a main airframe contractor. This contractor manufactures components in another factory to both the close tolerances and high quality assurance level required. The secondary airframe components are subcontracted by the main airframe contractor to yet another manufacturer whose normal business is the manufacture of airframe parts for commercial subsonic aircraft. They are provided with the necessary high-strength material 'free issue' along with instructions for the special forming and joining processes that are required.

The various hydraulic systems are grouped together within a single contract which is let directly by SEI to a specialist hydraulics subcontractor. This contract is let on the basis of a simplified functional specification, i.e. stating in abridged terms only what the final systems are *required to do*. The specialist subcontractor then does the detailed design and assembles the systems from standard components. The units are functionally tested in the works, using the computer systems supplied by SEI, complete with the necessary software.

Some important and fundamental concepts are relevant to this case study. They have relevance also, in a generic form, to all complex projects involving technology transfer. There are eight of them, as follows.

Design levels

For almost any engineering product, the design activity must be divided into at least two levels: *conceptual* design and *detailed* design. Conceptual design involves basic decisions about the function of an item, the principles on which it works, and broad operating parameters such as speeds, pressures and factors of safety. Conceptual design is often done as part of the early work (known formally as 'engineering' work) carried out by a design contractor. Detailed design is the more labour-intensive process which follows — to transform the conceptual design into a full set of engineering drawings and corresponding technical specifications. Detailed design is generally done by the company who will manufacture the item to the conceptual design criteria that have been set.

Design control

Design control is the term given to the activity of *controlling* a design between the point of conception and the completion of the finished, assembled product. It includes such activities as controlling and recording design parameters like materials sizes, inputs/outputs and various activities involving the design being received by other parties.

Change control

This refers to changes, not only to design parameters, but also to other aspects such as drawings and detailed work instructions. Change control of the minutiae of design and manufacture is one of the activities that contributes significantly to a precisely engineered product.

Project structure

You can think of project structure as being exclusively about management responsibility rather than anything technical. A good structure defines *who is responsible* for managing the component parts of a project and infers the routes that reports and other communications should take.

Contract compliance

It is one thing to give a manufacturer a contract to produce a specified product but quite another to receive the completed product absolutely compliant with your requirements in all respects. Errors in specification, interpretation and implementation, with the best will of all parties, occur frequently. An essential part of any design or manufacturing contract therefore is the activity of *checking compliance*. In practice this involves reviewing documents, assessing manufacturing capability at several stages, and then checking the final specification and quality of the finished article.

The supply chain

Industrial products are very rarely produced by a single manufacturer. Sub-assemblies, standard components and often basic materials are procured via a hierarchy of commercial supply contracts. Each contract, in turn, normally contains its own hierarchy of subordinated responsibility. This generic arrangement, which is universal, is termed the *supply chain*.

Technical interfaces

At the upper levels of project structure, decisions are made about the way in which the design and manufacturing contracts will be subdivided. Each contract has a physical interface with 'adjacent' contracts: the point at which the equipment supplied under each scope of supply joins together. This is known at the practical level as the *contract interface* and each contract may have several. Conceptually, the apportioning of responsibilities tends to be referred to as *technical interfacing* – although this has more management implications than is suggested by the title. In practice, you will often see the terms 'contract interface' and 'technical interface' used synonymously. This simplifies matters, as long as you understand the context in which it is used, and therefore understand the nature of the interface which is actually being referred to.

Quality management

This is a much used (and abused) term. In essence, the term 'quality management' relates to a system of paper procedures that form the *documentation part* of controlling what happens during the design and manufacture of a product. There is a link, of sorts, with the inspection of equipment during manufacture, and with the fitness for purpose of the finished item. The most commonly accepted international reference document in this area is the ISO 9000 series. It consists of four main parts:

- ISO 9000–1 (1994) *Guidelines for selection and use*
- ISO 9002 (1994) *Quality systems, model for quality assurance in production, installation and servicing*
- ISO 9003 (1994) *Quality systems, model for quality assurance in final inspection and test*
- ISO 9004–1 (1994) *Quality management and quality system elements – guidelines.*

These standards can be tailored to almost any application – but only because they are so broad as to allow significant room for interpretation. A related standard, which deals more directly with the management of design, is:

- BS 7000, *Guide to managing product design.*

This document is a very useful complement to the ISO 9000 series. It follows an almost identical philosophy but relates the various terminology much more directly to *design* than ISO 9000 does.

11.4 The problem

What exactly *is* the problem with technology transfer projects? Do the main difficulties lie with the technology, or are they something to do with the transfer?

Most projects have problems with both of these – though often at a superficial level. The core problem, the *real* problem, is one of *complexity*. Complexity is the one thing that characterises all technology transfer design projects, because it is inherent in the complex web of activities that are needed to make the project work. A project that is capable of completion (and that is the object of the exercise) must, by definition, contain complexity.

So:

The core problem of technology transfer design projects is ... **COMPLEXITY**.

Fortunately, it is not that difficult to address problems of complexity. Complexity has an opposite – two in fact. The first is *organisation*, and the second is *structure*. Whereas each specific technology transfer project will have different characteristics, and so an overall set of prescriptive 'rules' would not be possible, organisation and structure are the lowest common denominators of good projects. Organisation and structure of what? This is a fair question. Here is the answer:

- *All parts* of a technology transfer design project need an actively thought out organisation and structure. You cannot leave it to chance (or even politics).

This is fine. All you have to do now is to address the content of a project – sticking to this philosophy of looking for lowest common denominator aspects of the problem. There are two essential parts: the management *aspects* and the technical *content*. Think about this for a moment – some form of management is an essential part, clearly, but so much of what actually happens in any type of design project has a technical root to it. Project management will only work if it is capable of being implemented within whatever real technical constraints are inherent in that specific project. This means that whilst organisation and structure aspects are generic to all projects, technical constraints are not – they are highly project-specific and always different.

Now we know the problem. The problem is how to structure (and organise, remember) technology transfer projects, but *only* in a way that will fit in with the 'systemic' technical constraints of the project.

What are the difficulties in doing this? There are several. The first (and the most fundamental) is the problem of *conception*. There are many facets, both technical and management-related, to a project – the territory is very wide – so it is hard to understand, to conceive it all. Secondly, on a technical level, these projects are involved and multidisciplinary (complexity again), and no-one can be a specialist in everything. This is the clear, acid reason why so many technology transfer projects fail. So now, not only do we know the problem, we also know how difficult it can be to solve. An answer lies in methodology. We will look at a methodology, and then see how it is applied to the *Piranha* technology transfer project.

An outline methodology

There are various steps or areas of activity associated with a technology transfer project. They are not totally discrete from each other (they do have some overlaps in timescale, for instance), but you *can look* at them as being broadly chronological. This will help your understanding – help bring a *structure* to bear (structure, remember, is one of the enemies of complexity). These steps can be grouped together, again broadly, so that they represent four *phases* of a project. In this aggregated form, the chronology of the phases breaks down a little, but does not present too much of a problem – the essential content is still there. The steps and phases are shown in Fig. 11.2. They can be summarised as follows.

Step 1: look at the conceptual design The term 'conceptual design' does not invite clear definition. You can simplify it by thinking of it as the overall *picture*

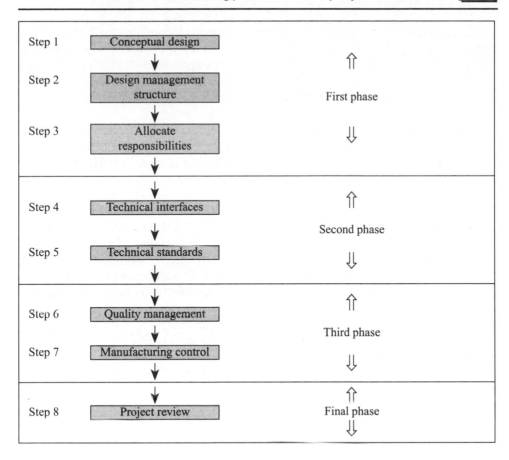

Figure 11.2 Project methodology – the steps and phases

of the systems (the engineering systems, not the management ones) that make up the engineering product – in this case the assembled rocket vehicle. Taken together, this assembly of engineering systems defines the overall extent of design activity that will take place.

Step 2: assess the design management structure   The *design* management structure is the way that the technological input is divided up between the participants in the project. Normally these mirror the split of technical responsibilities that have been agreed by the different companies involved in the project. Note the emphasis here on the *design* role rather than anything to do with formal project management responsibility as such. The partner company owning and providing the core technology can be considered, conceptually, as near the centre of the design management structure. You could think of this structure as a network of 'boxes of design responsibility'.

Step 3: allocate responsibilities   This is where you run into the question of project *management* responsibilities, but perhaps not in its purest form. Despite the nature of technology transfer projects – some of them almost attain the status of a global co-operative – they do tend to contain separate hierarchies for technical and management responsibility. In fact it is not uncommon for there to be a clear split between technical responsibility and the project management role, although you may have to look quite hard to see it. The secret of allocating responsibilities is to do so in a way that is systemic with this technical split – so it *fits in with it*.

Step 4: define the technical interfaces   A technical interface is a point in the design and manufacturing cycle where the technical impact of a company stops, and that of another co-operating company in the project begins. This interface really is purely technical – it is points on a drawing, or items on a parts list. Try not to think of it as being linked to commercial issues; that may be the reality but it won't help you think about technology transfer project structure in the right way. Leave it to others. Returning to this technical level, you can look for technical interfaces at two main levels, this time with a very definite chronological link. *Design interfaces* should be decided early in a project; they basically decide who designs what. A design interface is a point at which relevant pieces of a design are intended to fit together. *Manufacturing interfaces*, strictly, only take effect after both the conceptual and detailed design of an assembly or system is finished. They are more tangible, and therefore simpler. Once drawings and parts lists are complete, then manufacturing responsibilities fall easily into place, guided by the engineering capabilities of the manufacturing companies involved in the project, either as part of a joint venture/ partnership agreement or as a purely commercial subcontract.

Step 5: specify the technical standards   This means what it says. Standardisation is an absolute necessity for projects in which technical activities are dispersed between several companies, often in different countries. It is also an important mechanism for technical control within the 'supply chain' discussed earlier. The level of technical detail of this case study, however, is such that the definition of technical standards is not a major requirement.

Step 6: manage quality   We discussed earlier the basic concept of *quality management*. At our current 'structural level' of analysis it is sufficient to say that the standard series ISO 9000 must play its part in the project. The reason, as before, is simple; standards provide structure, and structure is the opposite of complexity. A project-wide policy of quality management, using the provisions of the ISO 9000 series, is therefore one of the cornerstones of effective technology transfer.

Step 7: control manufacture   This step starts at the beginning of manufacture of the components and systems and ends *only* when the parts are complete, assembled and correct. The activity of checking compliance (introduced earlier) is a big part of this step. Note that the real issue is one of control, rather than just monitoring. Fortunately these activities are included in the philosophy of ISO 9000, and product conformity (compliance with specified requirements) is a central tenet of the standard.

Step 8: the project review   The project review is the final act. It is the inquest into the previous seven steps. In theory, if you do it properly, you won't make the same mistakes twice. Looking back at Fig. 11.2, you can see how these eight steps are divided up into the four basic *phases* of a technology transfer project. This case study will look at each of the four phases but in varying levels of detail. Phase 3 involves a greater degree of structural and organisational detail than the others and you should now see how the basic concepts outlined earlier in the text fit into the various steps.

11.5 Case study tasks

The case study task is divided into two parts. Task 1 is an exercise in developing the structure of a technology transfer project – so you can learn how to deal with complexity. Task 2 is more concerned with organisation, through the specific mechanism of the quality management standard ISO 9000.

Task 1: project structure

Firstly, read the text of the case study again: the background information, the technical details of the space vehicle construction and the contract outlines. Try to understand the basic concepts in the context of the way that they fit into the eight project steps. Then, using what you have read, attempt the following:

* Draw a two-dimensional 'block diagram' of the project structure. This should incorporate, within the conceptual design of the space probe and rocket vehicle, the design management structure (step 2) and the allocation of responsibilities (step 3).
* Annotate the diagram showing the location of the *technical* interfaces – then write a simple description (no more than two lines) of each of the interfaces you have identified. Aim for between 10 and 15 of these interfaces.
* Answer the following question in 150–200 words. If, in the year 2002, you heard that the *Piranha* space vehicle project had been cancelled due to 'cost overruns' and 'unreasonable technical problems', what would you now be able to conclude?

Task 2: quality management elements

Your task is to prepare a quality management 'policy document' to be applied at the senior management level of the *Piranha* technology transfer project. To be suitable as a policy document for all the participating companies it needs to be quite short and to the point. It has to identify clearly, however, *specific quality management elements* to be used – it is no use making it too 'general'. Although the concept of the quality management system is that it should comply with ISO 9001 (as in real commercial practice) you may find it easier to refer to the following standards first:

* BS 7000: *Guide to managing product design*
* ISO 9004–1: *Quality management and quality system elements – guidelines*.

You will need to choose from the various quality system elements given in these two standards those that are most relevant to the specific problems of a complicated, diverse, technology transfer project. Use the eight project steps shown in the case text as a framework to work to.

References

11.1 BS 7000: *Design management systems.* (1989) Part 1 – *Guide to managing product design*; (1995) Part 3 and Part 10 – *Glossary of terms used in design management*. British Standards Institution, London.

12 Mechanical seals – improving design reliability

Keywords

Reliability in design (concepts) – MTBF/MTTF – the 'bathtub curve' – failure mode analysis (FMA) – risk analysis – design reliability improvement (eight principles) – mechanical seal design features and parameters.

12.1 Objectives

This case study addresses a difficult area of engineering – the connection (and sometimes conflict) between a theoretical and a practical approach. The subject of reliability in design is a good area in which to introduce this; reliability analysis is a well-developed subject, with a robust mathematical and theoretical basis. Equally valid, however, is the activity of *improving design reliability* using basic engineering principles. This is a simpler, more practical approach – it is also well supported by theory but it is more 'in the background' – not so obvious at first glance. The purpose of this case study is to provide you with an understanding of these two approaches, using as an example the design of a simple mechanical shaft seal. Mechanical seals are prone to failure from several causes and provide a good representative example of the fundamentals of reliability assessment and type of improvements that can be adapted to other, more complex, items of engineering equipment. The principles are the same.

12.2 Background: mechanical seals

Mechanical seals are used to seal either between two working fluids or to prevent leakage of a working fluid to the atmosphere past a rotating shaft. This rotary motion is a feature of mechanical seals – other types of seal are used for reciprocating shafts, or when all the components are stationary. Figure 12.1 shows a typical mechanical seal and Fig. 12.2 a specific design with its component pieces. They can work with a variety of fluids and, in the extreme, can seal against pressures of up to 500 bar, and have sliding speeds of more than 20 m/s. The core parts of the seal are the rotating 'floating' seal ring and the stationary seat. Both are made of wear-resistant materials and the floating ring is kept under axial force from a spring or bellows to force it into contact with the seat face. This is the most common type and is termed a 'face seal' – found in common use in many engineering applications, such as vehicle water

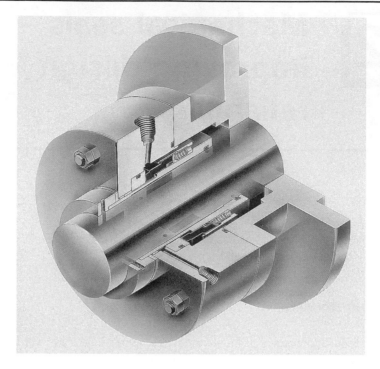

Figure 12.1 A typical mechanical seal (courtesy John Crane Mechanical Seals Ltd, UK)

pumps and automatic transmission gearboxes, washing machines and dishwashers, as well as more traditional industrial use on most types of process pumps. Materials of construction are quite varied, depending on the characteristics of the process fluid.

Mechanical seals are mass-produced items manufactured in a large range of sizes. Special designs are required for use with aggressive process fluids such as acids, alkalis and slurries. The design of mechanical seals is quite specialised and has developed iteratively over many years. Two points are important. First, seals are a good example of *closed design* – a concept introduced briefly in Chapter 2. This means that the external variables which affect the design may not be immediately obvious, and so have to be determined as part of the design process. Second, the design of a seal is a *multidisciplinary* activity, involving several different engineering disciplines. As with several of the other case studies in this book, the linking of these disciplines is a key part of the design process. Briefly, the engineering disciplines that are used in mechanical seal design are:

- *Thermodynamics*. Heat transfer in the seal components and the fluid film must be considered.
- *Fluid mechanics*. Hydrodynamic, boundary and static lubrication conditions exist in various areas of the sealing face and associated parts of the seal assembly. Laminar flow calculations govern the leakage path between the floating seal ring and stationary seat. Static fluid pressure considerations are used to determine the additional axial 'sealing force' generated by the process fluid.

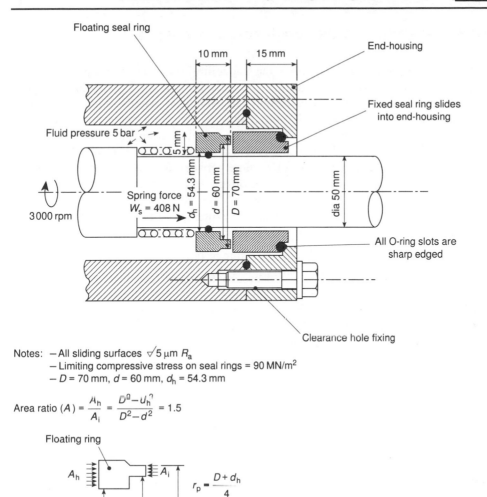

Figure 12.2 The basic mechanical seal

- *Deformable body mechanics*. Deformation of the seal ring in use is an important design parameter. This is calculated using classical 'hollow cylinder' assumptions with 'open-end' boundary conditions. Local deformations of the seal and seat faces also need to be considered.
- *Surface mechanics*. Surface characteristics, particularly the roughness profile, of the contacting faces affect leakage, friction and wear. This plays an important part in the tribology (the study of moving surfaces in contact) of the sealing faces.
- *Materials technology*. Material properties also play a part in the tribology of mechanical seals. The compatibility of the seal faces must be considered. Wear resistance and friction characteristics are influenced significantly by the choice of materials – many seals use very specialised wear-resistant materials such as plastics or ceramics.

12.3 Reliability in design – some concepts

How do you measure, or even define, the 'reliability' of an engineering component? Harder still, is it possible to quantify the reliability of an engineering design – perhaps before the component or assembly has even been manufactured? In practice, this scenario is very difficult; if you look, you can find increasingly complex definitions, some of them contradictory, as to what constitutes this quality of *reliability*. There is a well-developed theoretical side to it – quantities such as MTTF (mean time to failure) and MTBF (mean time between failures) are in common use in safety-critical applications such as petroleum and chemical plant design. Their purpose, however, is one of quantification only, i.e. to provide a common yardstick against which the concept of reliability can be expressed. It is not so certain that they will help you understand what reliability actually *is*. Consider this statement:

• Reliability is about *how, why* and *when* things fail.

From a design point of view it is still necessary to accept that there are two sides to this; the theoretical 'background' containing calculated lifetimes, MTTFs and MTBFs, but based on probabilities, and the practical engineering aspects, which do not always follow such well-defined rules. Try to accept that the two sides can co-exist. Using this as a background, we can now look at some of the practical concepts that can be of use to you.

The 'bathtub curve'

This is so-called for no better reason than that it looks, in outline, like a bathtub (see Fig. 12.3). There is no evidence to support any suggestion that the curve has anything to do with the reliability of bathtubs, large or small, anywhere. It indicates *when* you can expect things to fail. Surprisingly, it is well proven, reflecting reasonably accurately what happens to many engineering products. It tends to be most accurate for complex products that contain a large number of components – motor vehicles, domestic appliances and rotating equipment components are good examples. Looking at Fig. 12.3 you can see how the chances of failure are quite high in the early operational life of a product item; this is due to inherent defects or fundamental design errors in the product, or incorrect assembly of the multiple component parts. A progressive wear regime then takes over for the 'middle 75 per cent' of the product's life – the probability of failure here is low. As the lifetime progresses, the rate of deterioration increases, causing progressively higher chances of failure. Remember that although this is a generalised curve, the correlation with real engineering products is good.

Failure mode analysis

Failure mode analysis (FMA) is concerned with *how and why* failures occur. In contrast to the bathtub curve there is a strong product-specific bias to this technique, so generalised 'results' rarely have much validity. In theory, most engineered products will have a large number of possible ways that they can fail (termed *failure modes*). Practically, this reduces to three or four common types of failure, because of particular design parameters, distribution of stress, or similar. The technique of FMA

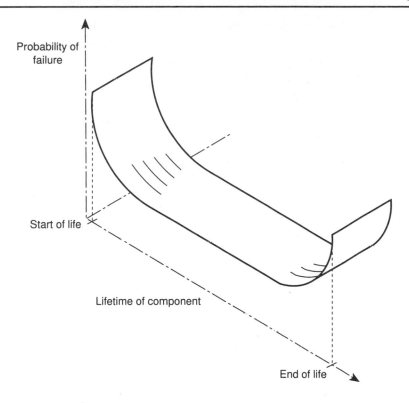

Figure 12.3 Component reliability – the 'bathtub curve'

is a structured look at all the possibilities, so that frequently occurring failure modes can be anticipated in advance of their occurring, and can be 'designed out'. FMA is therefore, by definition, multidisciplinary. Figure 12.4 outlines the principles of FMA, using as an example a simple compression spring – a common subcomponent of many engineering products. Bear in mind that as a spring is one single component, the FMA is simplified; things become more complicated for assemblies that contain many different pieces. The same principles, however, continue to apply.

Risk analysis

This is traditionally a loosely-used term. It encapsulates a number of assessment techniques which have in common the fact that they are all *probabilistic*. They look at a failure in terms of the probability that it may, or may not, happen. The techniques tend to be robust in the mathematical sense but sometimes have rather limited practical application. There is no doubt that the mathematics of probability is a sound science (much of atomic theory and wave/particle physics is built around it) but it does not have such an invariant effect as do, for instance, Newton's laws of motion. The 'rules of probability' are not axiomatic in the engineering world. This means that risk analysis techniques are fine, as long as you realise the limitations of their use when dealing with practical engineering designs. You should not need to use them in this case study.

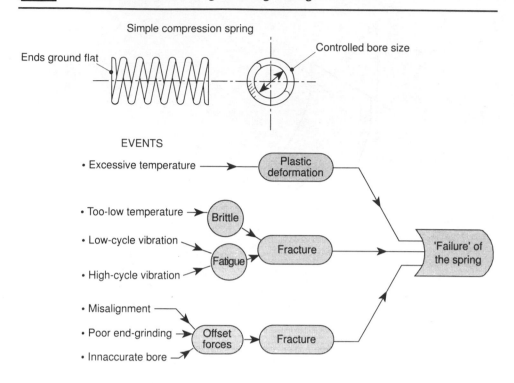

Figure 12.4 The principle of failure mode analysis (FMA)

Reliability assessment

A perfect assessment technique needs hindsight. It is not too difficult to decide why something has failed, once it has done so. The most useful form of reliability assessment involves *looking forwards,* to try and eliminate problems before they occur, to effect reliability improvements. There are limitations to this technique:

- It involves *anticipation*, which is difficult.
- Reliability assessment is strictly *relative*. It may be possible to conclude that component X should last longer than component Y, but not that component X will definitely last for 50 000 hours.
- It must be combined with sound engineering and *design knowledge* if it is to be effective.

Given these limitations we can now look at the methodology of design reliability assessment and think forward as to how to improve component reliability. It is based on a few sound and thankfully straightforward principles. It is easier to consider these in the general design sense in order to understand the concepts. They can then be adapted for specific product design cases.

Improving design reliability: eight principles

Reduce static loadings Reducing static loadings on component parts, by redistributing loads or increasing loaded areas, is good practice. The effects are small when existing design stresses are less than about 30 per cent of yield strength (R_e) but can be significant if they are higher. The amount of deformation of a component is reduced, which can lead to a decreased incidence of failure. Low stresses improve reliability.

Reduce dynamic loadings   In many components, stresses caused by dynamic loadings can be several orders of magnitude higher than 'static' design stresses – up to nine or ten times if shock loads are involved. Dynamic shock loads are therefore a major cause of failure. It is good design practice to eliminate as many shock loads as possible. This can be done by using design features such as damping, movement restricters, flexible materials and by isolating critical components from specific externally induced shock loadings. The possibility of general dynamic stresses can normally be reduced by reducing the relative speed of movement of components. In rotating shafts, particularly, this has the effect of reducing internal torsional stresses during starting and braking of the shaft.

Reduce cyclic effects   Cyclic fatigue is the biggest cause of failure of engineering components. High-speed, low-speed and normally static components frequently fail in this way. The mechanism is well known; cyclic stresses as low as 40 per cent of R_e will cause progressive weakening of most materials. One of the major principles of improving reliability therefore is to reduce cyclic effects *wherever possible*. This applies to the amplitude of the loading and to its frequency. Typical cyclic effects are:

- *Vibration*: this is defined in three orthogonal planes x, y and z. It is often caused by residual unbalance of rotating parts.
- *Pulsations*: often caused by pressure fluctuations.
- *Twisting*: in many designs, torsion is cyclic, rather than static. Heavy-duty pump shafts and engine crankshafts are good examples.
- *Deflections*: designs which have members which are intended to deflect in use invariably suffer from cyclic fatigue to some degree. In applications such as aircraft structures, fatigue life is the prime criterion that determines the useful life of the product.

Reduce operating temperatures   This applies to the majority of moving components that operate at above ambient temperature. Excessively high temperatures, for instance in bearings and similar 'contact' components, can easily cause failure. The principle of reliability improvement is to *increase the design margin* between the operating temperature of a bearing face or lubricant film and its maximum allowable temperature. Typical actions include increasing cooling capacity and lubricant flow, or reducing specific loadings. The effect of temperature on static components is also an important consideration – thermal expansion of components with complex geometry can be difficult to calculate accurately and can lead to unforeseen deflections, movements and 'locked in' assembly stresses. Stresses induced by thermal expansion of constrained components can be extremely high – capable of fracturing most engineering materials quite easily. For low-temperature components (normally static, such as aircraft external parts or cryogenic pressure vessels) the problem is the opposite: low temperatures increase the brittleness (decrease the Charpy impact resistance) of most materials. This is hard to 'design out' as low temperatures are more often the result of a component's environment, rather than its actual design. As a general principle, aim for component design temperatures as near to ambient as possible. It helps improve reliability.

Remove 'stress-raisers' Stress-raisers are sharp corners, grooves, notches or acute changes of section that cause stress concentrations under normal loadings. They can be found on both rotating and static components. The stress concentration factors of sharp corners and grooves are high, and difficult to determine accurately. Components that have failed predominantly by a fatigue mechanism are nearly always found to exhibit a *crack initiation point* – a sharp feature at which the crack has started and then progressed by a cyclic fatigue mechanism to failure. Techniques which can be employed to reduce stress-raisers are:

- Use *blended radii* instead of sharp corners, particularly in brittle components such as castings.
- For rotating components like drive shafts, keep rotating diameters as constant as possible. If it is essential to vary the shaft diameter, use a taper. Avoid sharp shoulders, grooves, keyways and slots.
- Avoid *rough surface finishes* on rotating components. A rough finish can act as a significant stress-raiser – the surface of a component is often furthest from its neutral axis and therefore subject to the highest level of stress.

Reduce friction Although friction is an essential part of many engineering designs, notably machines, it inevitably causes wear. It is good practice therefore to aim to reduce non-essential friction, using good lubrication practice and/or low-friction materials whenever possible. Lubrication practice is perhaps the most important one; aspects such as lubricating fluid circuit design, filtering, flow rates and flow characteristics can all have an effect on reliability. If you can keep friction under *controlled conditions*, you will improve reliability.

Design for accurate assembly Large numbers of engineering components and machines fail simply because they are not assembled properly. Precision rotating machines such as engines, gearboxes and turbines have closely specified running clearances and cannot tolerate much misalignment in assembly. This applies even more to smaller components such as bearings, couplings and seals. Reliability can be improved therefore by designing a component so that it can *only* be assembled accurately. This means using design features such as keys, locating lugs, splines, guides and locating pins which help parts assemble together accurately. It is also useful to incorporate additional measures to make it impossible to assemble components the wrong way round or back-to-front. Accurate assembly can definitely improve reliability (although you won't find a mathematical theory explaining why).

Isolate corrosive and erosive effects Frankly, this can be difficult. As a general principle, however, corrosive and erosive conditions, whether from a process fluid or the environment, are detrimental to most materials in some way. They cause failures. It is best to keep them isolated from close-fitting moving parts – the use of clean flushing water for slurry pump bearings and shaft seals is a good example. Corrosion and erosion also attack large unprotected static surfaces, so highly-resistant alloys or rubber/epoxy linings are often required. Galvanic corrosion is an important issue for small closely matched component parts of machines – look carefully at the electrochemical series for the materials being used to see which one will corrode sacrificially. Good design reliability is about making sure the less critical components corrode first. It is sometimes possible to change the properties of the electrolyte (this is often process or flushing fluid, or oil) to reduce its conductivity, if a potential difference exists between close-fitting components and galvanic corrosion problems are expected.

These eight principles form the core content of design reliability assessment and improvement. You can see how they are multidisciplinary and interlinked but they can be separated, as shown, to help you think clearly about the different elements of design reliability. The technique is to consider each one in turn, in relation to a specific product design, and look for individual changes (improvements) that can be made. A final *synthesis* will be necessary at the end to consider the totality of the suggested design changes, i.c. how they fit together. This synthesis is essential – if you don't do it you may make the design worse rather than better.

12.4 A closer look at seal design

Seal area ratio

In practice, most mechanical seals do not rely only on the force of the spring to keep the seal faces in contact (termed 'closure'). Closure is mainly achieved by the net hydraulic fluid pressure acting on the seal floating ring. This net hydraulic pressure is a function of the differential areas of the floating ring – hence the closure force increases as the sealed fluid pressure increases and the spring actually plays little part. Figure 12.5 shows how the closure force is made up of four components. The net hydraulic force (W_h) comes from the sealed fluid at pressure (p_i). This is joined by the spring force (W_s), the 'opening' force (W_o) caused by the seal interface fluid pressure and the friction force (W_f) caused by the frictional resistance of the static seal. Opening force (W_o) is normally calculated using the assumption that fluid pressure varies in a linear way across the radial seal face. The frictional resistance force (W_f) is just about indeterminable and can be ignored. Note the locations of the static 'O-ring' seals in Fig. 12.2 – they are an essential part of the seal assembly, to eliminate static leakage paths.

Seal ring dimensions

Mechanical seal design includes static calculations on the seal rings, which must have a sufficient factor of safety to avoid bursting. A valid assumption used is that face seal rings behave as hollow cylinders with open ends. Hence, for a hollow cylinder with internal radius (r), external radius (R) subject to internal pressure (p_i) and external pressure (P) it can be shown that:

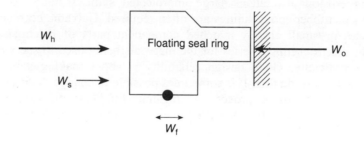

Net closure force $W = W_h + W_s - W_o \pm W_f$

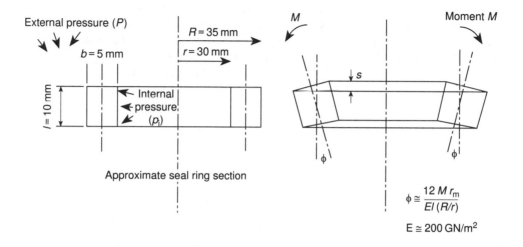

$$\phi \cong \frac{12\,M\,r_m}{EI\,(R/r)}$$

$E \cong 200\,\text{GN/m}^2$

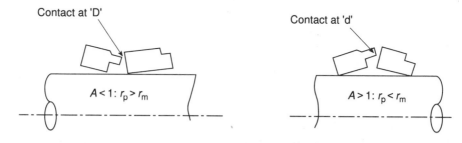

The effect of area ratio (A)

Figure 12.5 Seal ring behaviour

Maximum stress (Lamé): $\sigma = \dfrac{p_i(R^2 + r^2) - 2PR^2}{R^2 - r^2}$

For internal pressure (i.e. when $P = 0$) the maximum tensile stress (σ_z) at the ring bore is given by:

$$\sigma_z 5 \frac{p_i(R^2 + r^2)}{R^2 - r^2}$$

or, for external pressure (i.e. where $p_i = 0$), the maximum compressive stress (σ_b) at the bore is given by:

$$\sigma_b = \frac{2P}{1 - (r/R)^2}.$$

These equations only apply, strictly, to seal rings of plain rectangular cross-section. They can be used as an order of magnitude check, in which case significant factors of safety are included to allow for any uncertainties.

Seal ring deflections

A further important design criterion is the twisting moment which occurs in the floating seal ring. This can cause deflection (distortion) of the seal ring surface. From Figs 12.2 and 12.5:

D = external diameter of seal interface
d = internal diameter of seal interface
d_h = recess diameter
r_m = mean radius = $(D + d)/4$
r_p = torque arm radius
b = seal interface width

The moment arm is $M = \bar{P}(r_p - r_m)$ where $\bar{P} = APb$ and

$$r_p = \frac{D + d_h h}{4}$$

(d_h is shown in Fig. 12.2). We can see that as the area ratio A tends towards being greater or smaller than unity then the lever distance ($r_p - r_m$) gets larger, hence increasing the moment M. This moment produces angular distortion of the floating seal ring. Referring again to Fig. 12.5 the angular distortion (ϕ) is given by:

$$\phi = \frac{12Mr_m}{El^3(R/r)}$$

This results in a physical deflection (s) given by

$$s = \phi bC$$

where C is a shape factor (near unity) related to the section of the floating ring.
   The result of this is that the floating ring will contact at one end of its surface. As a general rule for rings under external pressure (as in Fig. 12.2):

- If $A < 1$, the floating ring twists 'inwards' towards the shaft (see Fig. 12.5) and hence contacts at its outer 'D' edge.
- If $A > 1$, the floating ring twists 'outwards' away from the shaft and hence contacts at its inner 'd' edge.

It is essential, therefore, when considering seal design, to calculate any likely twist of the floating seal ring to ensure that this is not so excessive that it reduces significantly the seal interface contact area. A maximum distortion (s) of 15 microns is normally used as a 'rule of thumb'.

Friction considerations

The main area of a mechanical seal where friction is an issue is the floating seal ring/seat interface. The whole purpose of this interface is to provide the main face sealing surface of the assembly with only a controlled degree of leakage, hence some friction at this face is inevitable. If it becomes too high, too much heat will be generated, which may cause excessive distortion of the components and eventual seizure. Given that rotational speed of the shaft is difficult to change (it is decided by the process requirements of the pump or machine) careful choice of the ring materials and their respective surface finish is the best way of keeping friction under control.

Surface finish is defined using the parameter R_a measured in microns. This is the average distance between the centreline of a surface's undulations and the extremes of the peaks and the troughs. It is sometimes referred to as the centreline average (CLA); see Fig. 12.6. The contacting surfaces of the floating seal ring and static seating ring must have closely controlled surface finishes – this is one of the core design criteria of mechanical seal design. If the surface is too rough, the effective contact area of the interface will be reduced, resulting in a significant increase in seal loading for a constant closure force (W). This can cause lubrication film breakdown and seizure. Experience shows that the lubrication regime existing between the interface surfaces is rarely completely hydrodynamic; boundary lubrication conditions provide a better assumption and these are prone to breakdown if specific contact loading is too high. Conversely, a surface which is too smooth is less able to 'hold' the lubricant film so whilst the effective contact area of the interface is increased with smooth surfaces, there may be negative effects on the stability of the lubrication regime. In practice it has been found that a surface roughness of 0.1 ± 0.025 μm R_a (both contacting rings) gives the best results.

In engineering terms, a surface roughness of < 1 μm is defined as a *precision finish* and requires a suitable manufacturing process. Standard turning on a lathe will rarely provide a finish better than 3.2 μm and even precision grinding can sometimes only achieve 2–3 μm, depending on the materials. Finishes of < 1 μm invariably require lapping, followed by machine polishing.

Assembly considerations

Many mechanical seals fail in the early stages of their life (the left-hand end of the bathtub curve) because of inaccurate assembly. This is normally due to one of two reasons.

Axial misalignment This is displacement of the seal end housing, and/or the static seat sealing ring so that its centreline lies at an angle to that of the rotating parts. This results in almost instantaneous wear and failure after only a few running hours. Misalignment can be prevented by incorporating features which give positive location of the seal end housing.

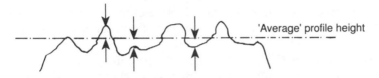

'Average' profile height

Surface roughness is described by R_a: the mean distance
of peaks and troughs from the 'average profile line'

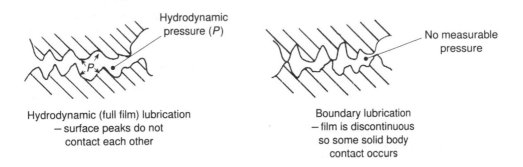

Hydrodynamic
pressure (P)

No measurable
pressure

Hydrodynamic (full film) lubrication
– surface peaks do not
contact each other

Boundary lubrication
– film is discontinuous
so some solid body
contact occurs

Figure 12.6 The effect of surface finish

Poor concentricity This is mainly a fault with the positioning of the fixed seal seating ring centreline, i.e. it is not concentric with that of the shaft. It can be almost eliminated by specifying the manufactured concentricity level of the components (BS308) and then providing a positive concentric location for the seal seating ring in the end housing. Note that this must still incorporate the O-ring, which prevents fluid leakage.

To avoid sharp changes in section of the rotating shaft, whilst still providing an abutment face for the closure spring, some seal designs incorporate a shaft sleeve. This fits concentrically over the shaft for slightly more than the length of the spring. The static O-ring rubber seals are often a problem during assembly as they can be chafed by the sharp edges of their slots. This causes the O-ring to lose its 100 per cent sealing capability.

12.5 Case study tasks

Your case study task consists of the reliability assessment and improvement of the design of a simple mechanical 'face' seal. This is shown in Fig. 12.2 – note that this figure is intended as a complete description of the design, i.e. you may assume that any engineering feature (material, dimension or tolerance) that is not shown in the figure has not been specified for this design. Many of the features and limitations of the design have been discussed in the case study text. Your task is to make changes to the design to *improve the design reliability* of the seal. Your suggested design changes should have a practical basis – they must be possible in engineering manufacturing terms, but they should also possess a sound theoretical basis, where possible. Divide your methodology into three steps:

- *List the design features*: parameters, dimensions, tolerances, etc., that contribute to the overall reliability of this seal design. It will help you if you can make this list as specific as possible.
- *Do a basic failure mode analysis (FMA)*: i.e. try to anticipate the various ways in which the seal could fail. There are various discussions about this scattered through the case study text, but you may also look at other references and proprietary seal manufacturers' catalogues.
- *Redesign the seal* for improved reliability. Using Fig. 12.2 as the starting point, incorporate the various design improvements which you feel are necessary. You should show these improvements by drawing a fully dimensioned sketch of the redesigned seal assembly, including all necessary information. Note that a properly scaled engineering drawing is not required – it should be a *design sketch*. To accompany the sketch, provide a short commentary (one or two paragraphs) on each design change, explaining the objectives and mentioning practical engineering aspects of the change.

12.6 Nomenclature

A	Seal area ratio
A_h	Hydraulic area
A_i	Interface area
b	Seal interface width
C	A 'shape factor'
d	Internal diameter of seal interface
d_h	Recess diameter
D	External diameter of seal interface
E	Young's modulus of seal ring
M	Moment arm
p_i	Sealed fluid pressure (internal)
P	Sealed fluid pressure (external)
r	Internal radius of a ring
r_m	Mean radius of seal interface
r_p	Torque arm radius
R	External radius of a ring
s	Deflection of seal ring
W_f	Friction force
W_h	Net hydraulic force
W_o	Opening force
W_s	Spring force
σ_b	Compressive stress at seal ring bore
σ_z	Tensile stress at seal ring bore
ϕ	Angular distortion of ring

13 Aircraft flight control – function and ergonomics

Keywords

The basics of flight – flight instrument principles – pressure instruments – gyroscopic instruments – errors. Instrument layout – functional groupings – ergonomic considerations – 'scanning'.

13.1 Objectives

Only a few tens of thousands of days have passed since the first heavier-than-air aircraft flew. Since then the speed of technological change in civil and military aircraft design has been quite dramatic – it is fair to conclude that modern aircraft represent one of the best examples of 'compacted' and fine-tuned engineering design.

To control an aeroplane in flight it is necessary to understand the principles of *dynamics* that influence the way the aeroplane moves and then link this with the human element – the pilot. Aircraft control is one of the most testing examples of this machine/human interface. Despite the predominance of dynamics in describing aircraft movement, the wider subject of flight *control* is highly multidisciplinary: a precision-tailored and ordered assembly of mechanical, static, dynamic and fluid mechanics principles. The objective of this case study is to tease out and identify these principles, and in doing so understand how they have been linked together to form a complex engineering system. You can find the overall concept, and outline methodology, of this approach explained in Chapter 2.

13.2 The basics of flight

Aeroplane flight is based on a straightforward system of forces and movements. In 'straight and level' flight the four primary forces of lift, weight, thrust and drag acting on the aeroplane are balanced, giving a state of dynamic equilibrium. The stability of the aeroplane is, in simplified terms, a function of the way in which these four forces act together. A key parameter of flight is an aeroplane's position in flight, i.e. whether the nose is pointing up, down, left or right. This is known as *attitude*.

In flight, an aeroplane has freedom of movement in three dimensions, i.e. six degrees of freedom. Three perpendicular axes are used as references (see Fig. 13.1) to help describe the motion. These axes all pass through the centre of gravity.

- Movement of the nose up or down, about the lateral axis, is termed *pitching*.
- Movement of the wing tips up and down, i.e. rotation about the longitudinal axis, is termed *rolling*.
- Movement of the nose left or right, i.e. rotation about the normal or 'directional' axis, is known as *yawing*. This obviously has an effect on the direction in which the aeroplane is pointing.

Attitude can be referred to any of these three axes. Simplistically, an aeroplane's three main control *surfaces* each control the primary movement about one of these axes.

- Pitch is controlled by the elevators: hinged sections of the tailplane.
- Roll is controlled by the ailerons: hinged sections of the wings.
- Yaw is controlled by the rudder.

In practice, however, things are not quite this simple and the movement of control surfaces also introduces secondary effects as follows.

- When an aeroplane *rolls* it will also tend to *yaw* towards the direction of the lower wing. This is due to airflow forces on the sides of the fuselage.
- Similarly, if a *yaw* is induced (using the rudder), the aircraft will also tend to *roll* towards the trailing wing, due to increased lift on the forward wing which is moving faster.

Figure 13.1 shows these effects in simplified form. Note that there are also some (less pronounced) effects on pitch attitude and airspeed as a result of these basic rolling and yawing manoeuvres.

Turning

Because of the interaction of combined primary and secondary effects, the task of *turning* an aeroplane is not quite as simple as it appears. The essential requirement of a turn is that the lift force acting on the aeroplane must be given a lateral (i.e. sideways) component – which is what causes the turn. This is done by inducing a *roll*, i.e. using the ailerons – then it is necessary to use the rudder to 'balance' the turn, counteracting the secondary yaw. If this is not done, the turn will be unbalanced, and the aeroplane is said to be *slipping into* the turn. If too much rudder is applied to compensate then the tail will tend to move outwards from the turn and the turn will be unbalanced in the other direction. This is known as skidding out. Slipping and skidding are equally undesirable as they make the turns difficult to control (Fig. 13.2).

Airspeed

Airspeed is the speed of the aeroplane relative to the airflow in which it is flying. Because of the effect of wind it is rarely equal to the aeroplane's speed relative to the ground (its groundspeed). The most commonly used term is true airspeed (TAS). The

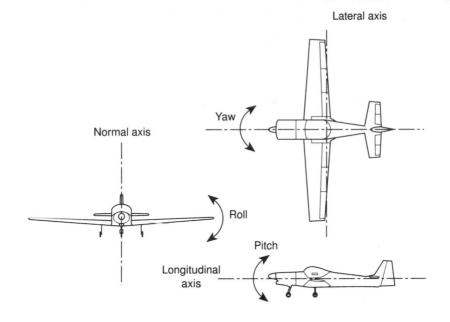

SECONDARY EFFECTS

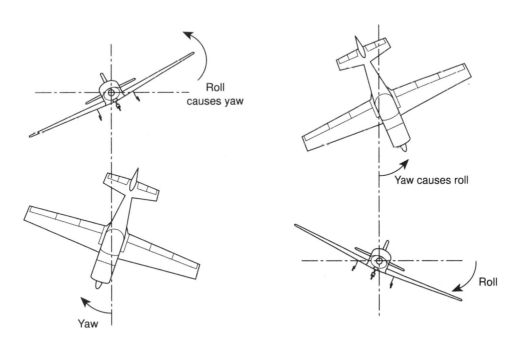

Figure 13.1 How an aeroplane moves

TAS at which an aeroplane travels is related to its engine speed (power) and attitude, i.e. whether it is climbing or descending.

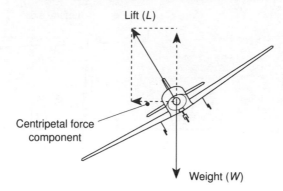

Turning: lift from the wings provides a centripetal component

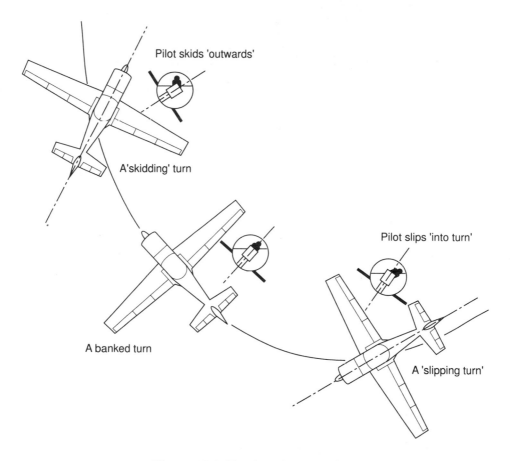

Figure 13.2 Turning the aeroplane

Direction

The direction of an aeroplane is considered in relation to fixed points on the earth's surface. Direction is controlled by turning the aeroplane as previously described.

Altitude

Altitude is the formal term for height. The most important altitude is of course that which is quoted with reference to the earth's surface. In practice the height of local terrain can change significantly over the earth's surface and so height is most commonly referred to a nominal ground level at which the atmospheric pressure is a constant value of 1013 mbar. This is then known as 'pressure altitude'. An aeroplane's altitude should not be confused therefore with its ground clearance, which is the straightforward vertical separation between the aeroplane and the ground. Altitude is an important parameter in the flight path of an aeroplane, whether it is climbing, descending or in level flight between destinations ('the cruise').

13.3 Flight instruments: the principles

The human senses, in their unassisted form, are of limited use when given the task of piloting an aircraft. Speed in the horizontal and vertical planes is difficult to judge when at a height of more than a few hundred metres from the ground and a sense of direction can only be maintained by reference to physical landmarks on the ground or in the sky (sun, moon, etc.). If the capacity for vision is lost, when flying in cloud for instance, then the human brain and body find it almost impossible to sense properly the attitude of the aircraft. For some reason, the brain seems to conspire to produce confusing signals – movements of pitch, yaw, roll and turning are difficult to tell apart. Flight instruments are therefore an absolute necessity for the safe control of an aircraft. They provide essential *feedback*. Because of the importance of their task it is possible to set down a number of overall requirements of the flight instrument set.

- Taken together, they must give a complete picture of what the aircraft is doing.
- Individual instruments must not be capable of contradicting each other – in fact it is best if they are able to *confirm* each other's readings.
- An individual instrument's ability to measure and display a particular form of movement or position must not be affected by other (different) aircraft movements.
- To be reliable, the instruments need to work using well-understood fundamental engineering principles. These principles need to be carefully chosen – the best ones are those that have minimum susceptibility to various types of errors.

The objective of flight control system design is to mesh together these essential requirements whilst bearing in mind the nature of the instrument/human interface, i.e. the *ergonomics*. Simplicity is important – the possible combinations of instruments are almost limitless. This means that it is necessary to consciously employ some simplifying influences. The main one is the concept of a *dominant instrument*. This is an instrument which, whilst it cannot itself describe fully what the aircraft is doing, provides the best first approximation. Such a concept also has ergonomic benefits; it is easier for a pilot to concentrate on one instrument than to divide his attention between several. Another important one is *functional grouping*. In simple terms this means that aircraft movements which have some similarity to each other are best displayed either on the same instrument, or on adjoining instruments. This helps the pilot's understanding (sometimes referred to as 'cognition') of what his instruments are telling him. With some principles established, we can now look at the individual instruments.

There are six separate instruments which comprise the main flight instrument 'set' displayed on the instrument panel of an aeroplane. They are divided into two broad categories: pressure instruments, which work using fluid (air) pressures, and gyroscopic instruments which utilise the dynamic principles of a spinning gyroscope. Most of the instruments also involve some type of mechanical/electrical interface. The choice of instrument functionality – what each one does – is allied closely to the basic mechanisms of flight control discussed previously. We can look individually at each of the flight instruments.

The attitude indicator (AI)

This is sometimes referred to as the 'artificial horizon' or the 'gyro horizon'. The AI has a circular display consisting of a representation of the 'real' horizon and a symbolic view 'from the aeroplane'. Its fundamental purpose is to show a simulated view of the aeroplane's attitude relative to the real horizon. It does this by showing pitch attitude and angle of bank (see Fig. 13.3). Note that the AI gives no information at all about the *performance* of the aeroplane, i.e. the way in which it is flying; the AI would show pitch and bank attitude if the aeroplane was simply being tilted about its lateral or longitudinal axes whilst stationary on the ground. It also provides no information on yaw.

The AI works using a horizontally spinning gyroscope with a vertical spin axis. This axis maintains its position relative to the earth's axis whilst the aeroplane effectively 'moves around' the gyroscope. These relative movements are displayed on the instrument as changes in pitch and bank attitude, as shown in the figure. Apart from cross-connections with the aeroplane's vacuum system, which makes the gyroscope rotate, the AI is not physically connected with any of the other five flight instruments. The AI suffers from various errors associated with gyroscopic instruments. Visual interpretation of the instrument is quite easy, the display being designed to mimic the view seen from the aeroplane cockpit.

The turn and balance co-ordinator (TBC)

Owing to the complexity of the turning manoeuvre this is a composite instrument having two separate (but related) functions: the turn indicator and the balance co-ordinator (refer to Fig. 13.3). The turn indicator consists of an aeroplane symbol which 'tilts' to the left or right. Its fundamental purpose is to show the *rate of movement* of the aeroplane about the normal axis – note that although it shows when the aeroplane is turning it does not indicate the angle of bank. It does not give any information about pitch or pure yaw either. This is a gyroscopic instrument utilising a single 'rate gyro' wheel with a horizontal (lateral) spin axis but it also has freedom to move about the aeroplane's longitudinal axis as shown in the figure. When the aeroplane turns, this causes a force to be transmitted to the gyro causing it to tilt. The tilting movement acts on a spring which causes the gyro to 'precess' with the aeroplane turn until the rates match. The amount of tilt is transferred by mechanical

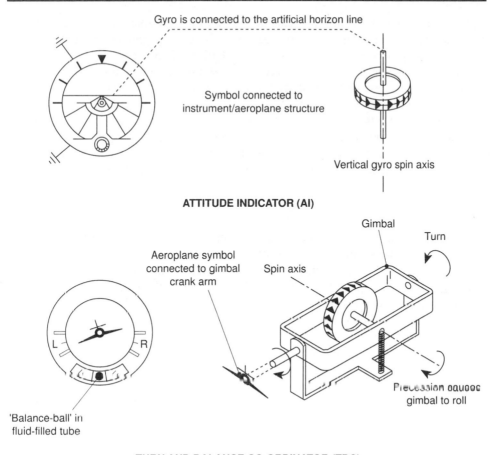

Gyro is connected to the artificial horizon line

Symbol connected to
instrument/aeroplane structure

Vertical gyro spin axis

ATTITUDE INDICATOR (AI)

Aeroplane symbol
connected to gimbal
crank arm

Spin axis

Gimbal

Turn

Precession oquooc
gimbal to roll

'Balance-ball' in
fluid-filled tube

TURN AND BALANCE CO-ORDINATOR (TBC)

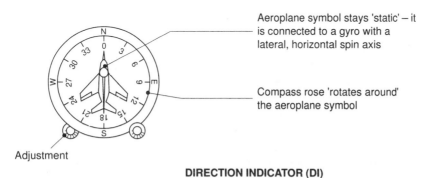

Aeroplane symbol stays 'static' – it
is connected to a gyro with a
lateral, horizontal spin axis

Compass rose 'rotates around'
the aeroplane symbol

Adjustment

DIRECTION INDICATOR (DI)

Figure 13.3 Gyroscopic instruments

linkage to cause a corresponding tilt of the aeroplane symbol on the instrument
display. The tilting aeroplane display is simple, but can be a little misleading –
remember that the tilt indicates *turning* rather than being an accurate indication of the

angle of bank. The turn indicator cannot provide full information about a turn and must be read in conjunction with the balance co-ordinator part of the instrument.

The balance co-ordinator consists of a small ball in a fluid-filled curved glass tube. Its purpose is to identify yawing forces, i.e. the movement of the ball will show whether the aeroplane is in a balanced turn, is 'slipping in' or 'skidding out'. Lack of balance can then be corrected by the pilot, using the rudder and ailerons to return the aeroplane to a balanced turn. This part of the instrument is purely a mechanical 'pendulum' indicator of balance and is unrelated to any of the gyros. The turn and balance indicators are normally combined in a single instrument display as shown. They are used together to provide the pilot with information about a turn.

The direction indicator (DI)

The most basic directional instrument used in an aeroplane is the magnetic compass. This suffers serious limitations, however, both as a result of errors due to acceleration and turbulence (when it becomes unsteady and difficult to read) and to magnetic variation, depending on its position relative to the earth's axis. For these reasons it is used as a back-up instrument only and does not play an active part in navigation. So how is the direction of an aircraft determined?

The direction indicator (DI), sometimes called the 'direction gyro', is a gyroscopic instrument. Although it can be aligned with the magnetic compass in flight, it operates independently from it and is therefore not subject to acceleration or 'magnetic' errors. The display uses a simple aeroplane symbol with the periphery of the dial being subdivided into 360° (see Fig. 13.3). Although the fundamental principle of a gyroscope is that it will maintain its alignment in space, various inherent 'gyroscopic errors' tend to cause the DI to 'drift'. It must therefore be re-aligned periodically with the magnetic compass during a flight.

The altimeter

The altimeter is a pressure-operated instrument. It resembles a simple anaeroid barometer except that it is graduated in units of distance, normally metres or feet. It works by measuring air pressure, utilising the fact that the air pressure of the atmosphere decreases by approximately 1 millibar for each 10 metres in height. The pressure measured is *static* pressure and so is independent of the airspeed of the aeroplane. The principle of operation is that a compressible, sealed metal canister contains a fixed amount of air. When the external air pressure reduces (as the aeroplane's altitude increases), the canister expands. This movement is transmitted via a mechanical linkage to an instrument dial (see Fig. 13.4). The dial normally has two pointers with units of 10^3 and 10^2 metres (or feet) to make it easy to read quickly.

For normal aviation the altimeter is zeroed using the international standard atmosphere (ISA) which assumes a mean sea level pressure of 1013.2 millibars. The altimeter is calibrated to read zero altitude at this external pressure and aircraft height above this level is formally known as 'pressure altitude'. As a pressure-operated instrument, the altimeter is sensitive to any errors in the 'static' pressure sensing point and to mechanical inertia delays in the anaeroid bellows.

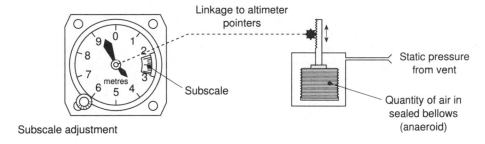

ALTIMETER

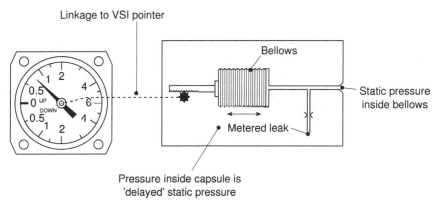

VERTICAL SPEED INDICATOR (VSI)

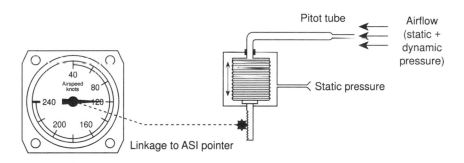

AIRSPEED INDICATOR (ASI)

Figure 13.4 Pressure instruments

The vertical speed indicator (VSI)

The VSI resembles a simple clock face but only has a single pointer which reads zero at the nine o'clock position (see Fig. 13.4). It provides the pilot with a direct readout of *rate of change* of altitude but provides no information about the actual height of the aeroplane above the ground – other than whether it is climbing or descending. The VSI works on the same principles as the altimeter, i.e. that air pressure decreases with altitude. It has a similar expandable canister but with a capillary tube leading back to the surrounding atmosphere. This causes a delay in the time it takes the canister to adopt its 'natural size' for the pressure altitude it is at and hence provides a 'trend' reading indicating whether the aeroplane is climbing or descending. This principle is shown in Fig. 13.4, although the actual mechanisms used are rather more complicated.

The airspeed indicator (ASI)

It was mentioned earlier that true airspeed (TAS) is the term used for the actual speed of the aircraft relative to the body of air in which it is flying. Unfortunately it is not possible to measure TAS easily. The nearest direct reading available to an aeroplane pilot is the parameter known as indicated air speed (IAS) which is measured by the ASI. The difference between TAS and IAS is caused by changes in air density. The ASI is a pressure-operated instrument. It senses the difference between the total pressure measured at a pitot-static tube and the static pressure measured at a 'static measuring point', where there is no dynamic component due to air velocity. This principle, and the instrument display, are shown in Fig. 13.4. The differential pressure reading is then transferred via a simple geared linkage to the single-pointer display. The display is calibrated in knots or kilometres per hour and is often colour-coded to indicate safe operating speeds of the aeroplane in various flight configurations; i.e. flaps up, flaps down, etc. Remember that these are all relevant to IAS, not TAS, which is *the* important parameter for navigation purposes. TAS normally exceeds IAS – the greater the altitude the greater the difference.

13.4 The problem: instrument errors

The design of aircraft instrumentation is beset with the problem of errors. In practice the various errors hinted at in the instrument descriptions cause significant problems with the instruments' operation in an aeroplane unless specific design measures are taken to minimise them. In most cases, the errors cannot be eliminated completely – the objective becomes one of minimising these errors so that they become manageable. The errors are the result of the fundamental engineering principles on which the instruments work. The two basic categories of instrument, i.e. gyroscopic or pressure operated, bring with them other undesirable properties that cause the errors. These are entirely multidisciplinary but can be broadly described as being due to properties of statics, dynamics, fluid mechanics and thermodynamics.

Gyroscope errors

Although gyroscopes provide a reliable way of determining the orientation of an aeroplane in space, they rely, for their accuracy, on precision manufacture. Small variations in component sizes due to thermal expansion can cause inaccuracies.

Gyroscopes use the principle of *precession* to 'move' the various parts of the instrument but this can be subject to errors caused by 'stray precessions' – a result of the instrument's orientation with respect to the earth. A similar effect is caused by acceleration of the aeroplane containing the instrument – the axis of the gyro becomes displaced, only returning to its correct position some time after the acceleration stops. This effect is particularly pronounced in 'rate gyros' such as those used in the turn indicator instrument. Gyroscopic instruments also suffer from friction effects – very fine manufacturing tolerances are necessary to ensure that the instrument remains reliable in use. The most drastic problem with gyroscopic instruments is 'toppling' – a condition in which the rotating gyroscopic assembly becomes displaced from its supports, due to the aeroplane performing violent manoeuvres such as steep turns or inverted flight. This can put the instrument out of action.

Pressure measurement errors

One of the main problems with measuring pressure in aircraft flight is the variation of air density with height. This affects the measurement of total pressure because of the effect that density has on the dynamic pressure component. Static pressure is measured using a vent on the surface of the aeroplane, or the static vent part of a pitot-static head. Total pressure is measured using a pitot tube mounted on the aeroplane skin, normally in the leading edge of the wing, where the airflow is least disturbed. The 'total' pitot pressure is fed via a system of small bore pipework to the ASI, altimeter and VSI. Air temperature reductions at high altitudes are known to have an effect on those instruments that operate on anaeroid principles, i.e. using a sealed bellows full of air. This is due to the gas laws and empirical results follow closely the results of theoretical analyses.

Pressure instruments that utilise expandable bellows suffer from inertia effects due to the finite time that a bellows takes to move through a set distance as a result of small pressure changes. Because of the sensitivity of the bellows design, even small pressure changes can be considered 'instantaneous' as a first approximation. Temperature effects on the bellows material and the mechanical linkages can generally be considered as 'second-order'. Instruments that use metered air leaks operate quite well in aircraft – the very small bore pipework, and particularly the small leak-orifices used, follow theoretical fluid mechanics principles well. Much smaller correction factors are required than in, for example, larger-diameter instrument piping systems used for land-based process plants and instrumentation. One important point is the pressure differential that exists across a 'metered leak'. As altitude increases, static air pressure (the low pressure side of the leak) reduces so that the leak flow characteristic is not necessarily constant – it may vary with altitude.

The issue of *cumulative error* is an important factor in aircraft flight instrument design. The objective is to try to make errors cancel each other out as far as possible. Compared to mechanically based errors, electronically based errors are considered by most instrument designers to be of second-order importance. Electrical design of aircraft systems is an interesting, but very different, subject.

13.5 Case study task 1

Chapter 2 of this book covered the basic steps of 'opening up' a design problem and then reducing what you find to a basic level before starting to *reconstruct* it in a way that you can understand. It is then easier to improve the design, or find the solution to a particular design problem in a way that will feel comfortable with. Task 1 in this case study follows these principles but concentrates on the pure technical aspects – the *results* of the methodology rather than the methodology itself.

Your task is to analyse the functional design of the aircraft flight instrument set with specific focus on the design aspects governing the *instrument errors*. A good starting point is to itemise, for each instrument in turn, the errors that are likely to affect the accuracy of that instrument, describing these in terms of fundamental engineering principles. The concept of 'root cause' is of key importance here – you are looking for the most basic engineering or physical law that is the absolute indivisible cause of the error. The best way to describe this is in a table. Figure 13.5 shows the principle using a partially completed example for the balance co-ordinator. Note though that this instrument is the one in the set that is least affected by errors – the others are more complicated. The right-hand column of the table is for conclusions that you draw about mechanical design parameters of the instrument, bearing in mind the 'error functions' identified. In a practical design review, this column would form the link between the functional design specification and the actual mechanical design criteria for the instrument – an important practical step. Note how the conclusions also include comments relevant to the pilot's interpretation of the instrument – we are starting to consider the man/machine interface and forming a link to Task 2 of this case study.

Instrument	Measured parameter(s)	Governing principle	Error observation	Definition of error	Design implications
Balance co-ordinator	Yaw about normal axis	Simple pendulum	Second order time-lag	Fluid friction only	Low-viscosity fluid used

Figure 13.5 Instrument errors – a starting point

13.6 Flight instrument layout

Introduction

The second part of this case study is about the physical layout of the six main flight instruments on the control panel of the aeroplane – an important consideration because the task of flying an aeroplane often involves information overload. This means that the safety and controllability of the aeroplane is restricted not by the limitations of the installed flight systems and instruments but by the sensory limitations of the pilot – high-speed military aircraft provide the most extreme example. The ease with which a pilot can interface with the flight instruments is therefore a key design consideration. To help the situation, flight instrument layout is designed to follow basic principles of *ergonomics*. The main concern here is *visual ergonomics,* rather than the layout of the flight control levers.

Four basic ergonomic principles have an input into flight instrument layout.

Principle 1: the dominant instrument

Sensory overload is most easily minimised by having a *dominant instrument,* centrally located on the instrument panel. This should be the instrument that provides, on a single display, the most extensive, salient information about the aeroplane's attitude. In terms of hierarchy of importance, the (safe) flight attitude of an aeroplane is more important than navigational aspects, i.e. where it is. In flight, the dominant flight instrument will be referred to more often than the other five.

Principle 2: functional grouping

Functional grouping means that those instruments which give essential complementary information about a flight manoeuvre should be located *next to each other,* or as near as is practical. As a guide, the manoeuvres that have to be considered can be thought of as being of three essential groups:

- climbing/descending
- turning
- height, speed, and direction.

Note how the third group is a composite – and does not relate directly to a predetermined manoeuvre. A good system of functional grouping will assist the pilot in assimilating quickly the information from the several instruments. This gives a good picture of how the aeroplane is behaving in, for example, a turn. The effectiveness of a functional grouping is nearly always a compromise – the objective of ergonomic design is to achieve the best compromise possible.

Principle 3: vision span

Humans find it easier to assimilate complementary information when the information displays are physically close to each other. Understanding, or *'cognition'*, is reduced when it is necessary to transfer attention consciously from one place to another. The distance between displays is termed the *vision span.* For full cognition the most effective vision span is dependent on the distance of the display from the observer's eyes. For a typical aircraft cockpit layout with an eye-to-display distance of 700–900 mm, it is approximately 120 mm; instrument displays spaced less than this distance apart can be assimilated together in a single glance, without the observer needing to transfer his gaze. Cognition reduces – approximately proportionally – as the spacing is increased. This is the fundamental reason why the turn indicator and balance co-ordinator are combined into a single instrument dial.

Principle 4: scanning

Scanning is the term given to the visual technique used by the pilot to assimilate the information provided collectively by the flight instrument displays. The technique is essential if the necessary amount of information is to be assimilated in the short time allowed by various flight manoeuvres. Scanning provides a structured rather than random way to do this. There are five rules to follow when designing a scanning technique: the technique must:

- reflect the relative importance of the dominant and supporting instrument(s);
- discourage omission of any instrument;
- avoid fixation of the pilot's gaze on a single instrument;
- recognise time lags and any other inherent errors;
- be adaptable to all the main flight manoeuvres: climbing/descending, turning and height/speed/direction changes.

Defining the scanning technique is an integral part of flight instrument panel design and layout. When a panel design is incorporated into an aircraft type, the documented scanning technique forms part of the operating instructions used in pilot training.

Figure 13.6 Cockpit layout – showing space for flight instruments
(courtesy Slingsby Aviation Ltd, UK)

13.7 Case study task 2

Your task is to design the layout of the six flight instruments in the cockpit display. The design should take into account the four visual ergonomic principles described. Figure 13.6 shows the overall cockpit layout and the physical space reserved for the flight instruments. Your design solution should:

- indicate the structured 'steps' that have contributed to the proposed solution;
- explain how the instrument set is used during the major flight manoeuvres described previously, i.e. climbing/descending, turning and height/speed/direction changes;
- show how the four visual ergonomics principles have been incorporated into the layout, and identify any major compromises that were necessary;
- include a written set of scanning technique 'rules' to be used during the major flight manoeuvres. They should be in the form of clear instructions to the pilot, but do not need to include detailed justification. You may wish to use the suggested matrix in Fig. 13.7 as a starting point.

Basic manoeuvre	Requirements to control the manoeuvre	Objective of scanning technique	Scanning instructions
Climbing	• AI to show slight 'pitch up' • Keep wings level (avoid bank) • Maintain minimum airspeed • Counteract yaw	To confirm the ASI reading with the pitch attitude indicated by the AI	AI – ASI – AI – ALT – AI – ASI – DI – AI – ASI – AI – TBC – AI

Figure 13.7 Instrument scanning – a starting point

13.8 Nomenclature

AI	Attitude indicator ('artificial horizon')
Ailerons	Hinged sections of the aircraft's wing, used to induce a roll
Altitude	Height above the earth's surface
ASI	Airspeed indicator
Attitude	An aircraft's position in flight, i.e. whether the nose is pointing up, down, left or right
DI	Direction indicator
Elevators	Hinged sections of the tailplane, used to induce a pitch up or down
IAS	Indicated airspeed – the airspeed reading shown by the aircraft's airspeed indicator (ASI)
Pitch	Movement of the aircraft's nose up or down
Roll	Rotation about the aircraft's longitudinal axis
Rudder	Hinged section of the tail fin, used to induce yaw
TAS	True airspeed – relative to the airflow in which the aircraft is flying
TBC	Turn and balance co-ordinator
VSI	Vertical speed indicator
Yaw	Movement of the aircraft's nose left or right

14 Power boilers – remnant life assessment

Keywords

Design lifetime – how things fail – power boilers, design and operation – high-temperature failures – low-temperature failures – creep and fatigue – fireside corrosion – gas side erosion – duplex failure mechanisms. Life assessment – life fraction rules – fatigue and creep life.

14.1 Objectives

Why don't engineered products last for ever? You may not often hear this question *asked* but it seems to be at the back of every purchaser's mind. Purchasers will, however, make no secret of the fact that the lifetime of a product that they buy is an important criterion in their buying decision. It is of little use having a product, however elegant and efficient, that will not give a reasonable working life.

Design lifetime is a term that you will see used in relation to engineered products. Its meaning is self-evident: the working life of the items, normally expressed in hours, from when it is commissioned to the point where it fails by some failure mechanism. This final point is important – items do not 'just fail' as such, it has to happen by a specific failure mechanism, such as creep, fatigue or similar. It is often necessary therefore to qualify statements of design lifetime by referring to an expected failure mechanism such as design *creep* life or design *fatigue* life. The other important term you will meet is *remnant life*. This refers to the amount of design lifetime remaining from the point at which an item is examined or assessed. It often has a safety basis; many items of engineering plant and equipment are subject to statutory (health and safety) legislation, requiring periodic survey and inspection during their life.

How things fail

Because the design lifetime of a component is closely linked to the question of how that component can fail, the issue of failure mode is of key importance. All the component parts of a product do not degrade at the same rate. A small number of components invariably 'age' much more quickly than the others, owing to the severity of the conditions that they see (temperature, stress, dynamic loading, etc.) until they become life-expired and fail. In many cases, the other components may be only partially worn, or in 'as new' condition. This brings into focus the question as to whether the design lifetime of a product can be controlled at the design stage. The answer to this has to be 'yes' – in many cases, particularly for large and expensive engineered products, design lifetime *can* be consciously designed into (or out of) the product by materials choice or design layout. Many smaller products such as vehicles and consumer goods are actively designed for a relatively short life, a concept known as planned obsolescence. Some manufacturers of low-priced consumer 'white goods' have turned this almost into an art form. The bottom line is twofold: the cost of the product and its profitability – there is no doubt that nearly all of the engineered products around you could be made to last longer than they do, but they would cost more; perhaps more than you would be willing to pay.

For these reasons, the issue of design lifetime is an important consideration in the design of engineered products. The case study will investigate this for a specific piece of equipment – a large power station boiler – and demonstrate how the various technical and engineering disciplines contribute to the overall concept of design lifetime. One of the technical objectives of this chapter is to show how calculations (at the initial design or later remnant life stage) are linked to an anticipated failure mode. These principles are common to many design cases and can also be used as part of failure investigation work.

14.2 The problem: remnant life of power boilers

The majority of electrical power in Europe is generated by fossil-fired (coal or oil) power stations utilising large steam boilers. These boilers can have a thermal output of up to 900 MW and are up to 80 m high. They constitute a huge capital investment, taking years to build. Modern boilers have a design life of approximately 20–25 years (175 000–200 000 hours) and provide a good example of time-dependent engineering design. The pressure parts of a boiler are subject to regular statutory inspection to check their condition – as the boiler gets older the extent of these inspections increases, and includes non-destructive testing (NDT) examinations of critical components. Because of the high capital cost of replacing a boiler it is common for several *life extension studies* to be carried out during its working life. At these stages the problem is how to *assess* the remnant life of the critical components. The results (and recommendations) must be robust enough so as not to compromise the safety of the boiler whilst taking into account the real financial pressures that exist for a boiler to remain in operation. Both must be considered against the technical complexity of the task of actually predicting when boiler components are likely to become dangerous, or fail.

14.3 Power boilers – design and operation

Power station boilers have evolved to their current large size and MW outputs over a period of some 30 years. There are approximately 50–60 of similar design in UK power stations, most of which are coal-fired. Boilers operate either 'base-load', for 24 hours per day, or a regime of 'two-shifting' where they supply power only for two periods per day (7 am and 5 pm) when demand is highest. A typical boiler has outlet steam conditions of 160 bar (gauge), 570°C and is of 'twin pass' design, in which the gas path is bent over at 180° to limit the physical height of the boiler (see Fig. 14.1). One pass contains a large vertical furnace lined with waterwall tubes – the burners being located in these walls, or at the corners, and fed with a combustible mixture of pulverised coal and air. The boiler is designed to produce superheated steam and has a reheat facility to restore heat to steam taken from the high-pressure (hp) turbine. There is an economiser stage located in the second pass. The gas temperature ranges from 1750°C in the furnace down to less than 300°C at the discharge of the second pass. The gas then passes through rotary air heaters where it transfers waste heat to air being fed to the boiler by the forced draught fans.

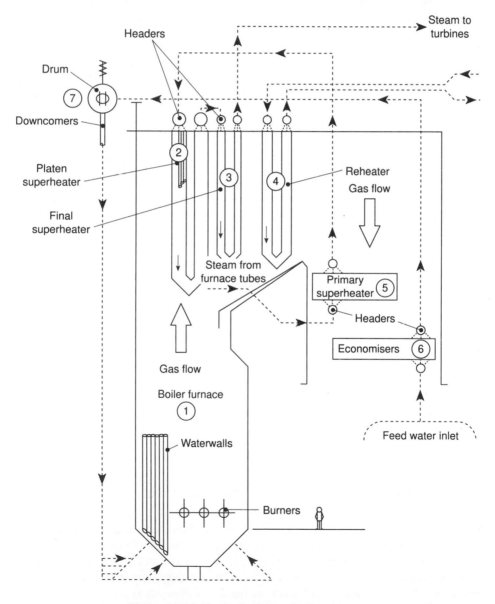

Figure 14.1 Typical power boiler layout

The components

The main components and the steam flow path are shown in Fig. 14.1. We can look at the location and function of the main components, taking them in order along the 'gas path'.

1. The furnace   The furnace comprises a rectangular area bounded by water-walls, vertical banks of 60–70 mm diameter low carbon steel (LCS) tubes joined by thin spacer strips. The pulverised coal burners are installed at the lower end. In operation, the furnace contains combustion flame and gases at temperatures of up to 1750°C. The circulating water within the waterwall tubes is heated predominantly by *radiation* to a condition of low-to-medium superheat. Radiant heat flux to the waterwalls is high, approximately 500 kW/m^2 in the hottest areas. The water circulates by natural convection as a result of density variations caused by the heating. Average metal temperatures in the furnace waterwall tubes are 270–300°C.

2. The platen superheater   At the top of the furnace, the gases pass first over the platen superheater. This is also known as the *radiant* superheater, as upwards of 80 per cent of the heat transfer to the water is obtained by radiation from the luminous combustion gases. The platen consists of banks of contraflow U-tubes hanging down into the furnace. The tubes are welded into thick-walled headers at their ends. The platen normally forms the second or intermediate stage of the superheater, heating the steam to approximately 500–520°C. Operating metal temperatures are high, up to 520°C, so heat-resistant alloys such as austenitic stainless steel need to be used.

3. The final superheater   This is of similar design to the platen superheater but smaller. It heats the steam to its final superheat condition (160 bar (gauge), 570°C) using mainly radiant, but also some convective, heat transfer. Headers are again used to connect the tube banks.

4. The reheater loop   The reheater accepts 'pass out' steam from the mid-stages of the generator turbine and reheats it back to near its original superheat condition. This increases the overall efficiency of the thermal cycle. The reheater loop heats the steam by mainly convective heat transfer from the furnace gases. Gas temperatures are approximately 1000°C in this area so heat-resistant austenitic alloys are required.

5. The primary superheater   This is the first stage of superheating, comprising a series of horizontal tube banks linked by headers. Owing to the lower temperatures in this region, ordinary low-carbon or C–Mn steels are used. The tube banks may be built up in several sections, the tubes being joined by circumferential welds.

6. The economisers   The economisers are the last stage of feed heating before the feedwater enters the boiler. They are horizontal banks of widely spaced (often finned) tubes located well down the second gas pass of the boiler. Gas temperatures drop to 280–300°C across the economiser banks. Construction is of low carbon steel.

7. The boiler drum The boiler drum is situated towards the top of the boiler and acts as the main steam/water reservoir. A typical drum will have dimensions of approximately 15 m length $\times$ 1.25 m diameter with a wall thickness of 40–50 mm (depending on the pressure and the construction standard). Water is carried from the lower half of the drum down into the boiler circulation system by large-diameter pipes called 'downcomers' (see Fig. 14.1). The drum operates at a temperature of 250–300°C and is of welded alloy steel construction. The inside of the drum contains steam separators, baffles, etc., known collectively as drum *furniture*.

14.4 Why boilers fail

Boilers are a good example of the need for thinking of failure modes in terms of their *priority*. Boiler components see a wide variety of different conditions and there are an equivalent number of deterioration and failure modes. The economics aspect is important when considering which modes should be given priority. Many of the smaller or easily accessible components of a boiler can be economically replaced and therefore are not a limiting factor in its economic life. Deterioration of some of the larger components will cause a situation where they cannot be replaced economically and they therefore act as 'life limiters'. Figure 14.2 gives an outline of the priority failure modes for the main boiler components – note that several different mechanisms are identified. We will now look at some basic definitions of failure mechanisms.

Boiler component	Metal temperature	Failure modes*
Furnace tubes	270–450°C	Fireside corrosion
Platen superheater	500–520°C	Tube fireside corrosion and wastage, **creep**
Final superheater	540–570°C	**Thermal fatigue** of headers and tubes. **Creep**
Reheater loop	540–560°C	Ash erosion, **corrosion fatigue**
Primary superheater	300–375°C	Ash erosion
Economiser	150–250°C	Ash erosion
Boiler drum	250–300°C	**Thermal fatigue**

* Life-limiting mechanisms are shown in bold type

Figure 14.2 Boiler components – failure modes

Low-temperature failures

The simplest, and most general, type of failure mechanisms are those occurring at low temperatures. Low temperature is generally taken as being below about 400°C, at which there is no time-dependent mechanism present such as creep. There are several types of low-temperature failures such as general yielding, buckling, fatigue and fracture. The category of 'fracture' is perhaps the most complex; you will often see it subdivided into three stages:

- *crack initiation*, where a small surface irregularity develops into a crack (by any of several methods);
- *crack growth*, in which a small crack grows to a size where it can cause a failure;
- *final fracture*, where the component breaks via a ductile/brittle fracture mechanism.

In practice, most parts of a boiler will fail by means other than low-temperature mechanisms. It should still be a consideration in any assessment though – it has important parallels to elements of high-temperature material behaviour.

High-temperature failures

High-temperature failures, those occurring above about 420°C, are predominantly due to the mechanism of *creep*. Creep occurs during steady running conditions and manifests itself in the *cracking behaviour* of a material. It causes microcracks and larger holes in the metal structure, causing it to lose its strength. The remaining material then suffers from increased stress loadings until it fails. There are several theoretical approaches used to describe creep mechanisms. Published creep/rupture data is available for high-temperature boiler steels such as $2\frac{1}{4}$Cr-1Mo – these are based on time/temperature duration either to failure or to achieve the initiation of a defect of a predetermined size. High-temperature headers are the main boiler components that fail by a creep mechanism. Damage is normally caused to the tube stub-welds in the header, or as a general swelling and distortion of the header body itself. The high-temperature superheater and reheater banks can also suffer from creep, causing vertical distortion of the tube bank and tube-to-header weld problems. High-temperature headers and superheater banks are difficult and uneconomic to replace towards the end of a boiler's life.

Thermal fatigue

Thermal fatigue resulting from heating/cooling cycle transient conditions can cause reduced lifetime of boiler components. Thermal cycling causes alternative tensile and compressive stresses in a component (the so-called 'hysteresis' loop) which exaggerate the stress conditions both at the surface of a component and in the root area of any small cracks and defects that exist within the material structure. Thermal fatigue has its effect therefore on the *crack growth* mechanism. As with creep, there are several theoretical models available to quantify cyclic fatigue and thermal cyclic fatigue mechanisms. Because temperature cycling is common to all parts of the boiler, many components may be subject to the effects of thermal fatigue. In practice, however, those components with heavy material sections such as headers, safety valve chests and, to a lesser degree, the boiler drum, are the worst affected. It is not uncommon for thermal fatigue to be the life-limiting mechanism for these parts.

Fireside corrosion

Corrosion of coal-fired boiler components is basically of two types: waterside corrosion and fireside corrosion. Waterside corrosion is rarely a factor affecting component life because feedwater chemical control is well understood and is given high priority during operation of the boiler throughout its life. The main problems are caused by fireside corrosion on those surfaces of the boiler tubes and other components exposed to the combustion gas. Fireside corrosion is caused predominantly by the fuel ash. Coal contains impurities in the form of alumino-silicates which form, during the combustion process, into ash. This ash contains products which have the general property of preventing (or 'unsettling') the process of stable Fe_3O_4 magnetite formation on the tube surfaces at their various operating temperatures. This prevents the tube being protected from the corrosive effects of the gas products and hence encourages corrosion. This commonly occurs in the following two places.

Furnace wall corrosion   The maximum metal temperature of the LCS furnace waterwall tubes (see Fig. 14.3) can be up to approximately 450°C under normal operating conditions. Although, theoretically, there is excess air in the furnace, practical considerations are such that local reducing conditions occur near the waterwalls, particularly as the boiler burner system gets older. The net effect of these reducing conditions is to prevent the formation of protective oxide scale on the tube surfaces – the scale that is formed is non-protective. Coal ash then settles on the tubes ('slagging') which further prevents the access of oxygen and makes things worse. One redeeming factor about furnace corrosion is that it does tend to be localised, often around the burner region and the lower end of the furnace.

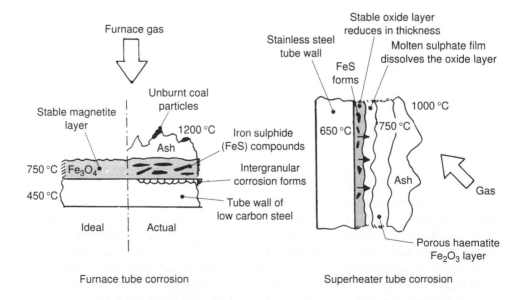

Furnace tube corrosion   Superheater tube corrosion

Figure 14.3 Differences between furnace and superheater corrosion

Superheater and reheater corrosion Corrosion of the superheater and reheater tube banks (situated at the top of the first gas-pass) presents a serious design problem. The radiant tube-banks suffer the worst, particularly those which are aerodynamically exposed to the full flow of furnace gases. They are made of austenitic stainless steel material (to resist creep) which is susceptible to ash corrosion, particularly by sulphate deposits which unsettle or restrict the formation of a protective oxide film. Corrosion often progresses to form 'wastage flats' either side of the tube. The extent of corrosion on superheater and reheater banks can be difficult to check, even with the boiler shut down, as they are difficult to reach. During operation, the surfaces are covered in molten slag and almost impossible to assess. This means that corrosion often continues undetected for some time.

The rate at which corrosion occurs is very difficult to quantify. It is a complex function of time, temperature, coal ash content and also physical and geometrical factors. Theoretical corrosion rates (in mm/year) are published for various materials but there is a large scatter associated with these data and their use tends to be limited to *comparisons* between materials, rather than specifying an absolute value which is reproducible in practice. Superheater and reheater tubes have quite thin walls and, although a design corrosion allowance is normally included in the wall thickness, failures are to be expected during the lifetime of the boiler.

Gas side erosion

Erosion of gas-side components occurs in coal-fired boilers because of the hard and abrasive (as opposed to corrosive) particles that exist in coal. Erosion is rarely a problem in the furnace and radiant areas where the high temperatures maintain the ash particles in a molten, soft condition. The main problems are experienced in the convective superheater and reheater banks where the temperatures have fallen below about 850°C, causing the particles to become hard and abrasive. Erosion rates are also a function of gas velocity; a design limit of 12–15 m/s is normally used. In practice, although erosion can be a serious problem, the components which are affected (the convective ones) can generally be economically replaced, so it is not often considered as a 'life-limiting' factor. Features such as sootblowing, where tubes are cleaned by jets of high-pressure steam to keep the tubes clean and therefore prevent increases in local velocities, are used to keep erosion to an acceptable level.

Stress-corrosion and corrosion-fatigue

These are two of the more important *duplex failure mechanisms* that exist. The concept of a duplex failure mechanism is an important one; although single, discrete mechanisms can be identified and described, most practical boiler failures are the result of a combination of mechanisms. Unfortunately it is not easy to incorporate this reality into design or remnant life assessments, other than to incorporate 'margins' into any quantitive assessments that are carried out.

Stress-corrosion is caused by the combination of normal working stress and a corrosion mechanism of some sort. Many alloys have a set of conditions under which they will suffer stress corrosion cracking (SCC) – particularly high-strength and stainless steel alloys. Fortunately, most boiler materials operate outside the rather narrow band of conditions under which SCC will occur, so SCC failures are usually random rather than generic. Those that do occur are nearly always related to poor design features such as crevices or stress-raiser points, both of which can stimulate SCC under the wrong conditions.

Corrosion-fatigue is a mechanism by which the fatigue life of a component (based on the traditional S-N curve) is reduced by the existence of a surface corrosion mechanism. Small surface defects, such as corrosion pits, cause cracks to be formed within fewer stress cycles. This allows crack growth and failure to occur much earlier in the life of a component than expected. There are three unfortunate features of corrosion fatigue:

- It is very *unpredictable*; it occurs in many different environments.
- It has a *synergy effect*; the damage caused by the duplex corrosion-fatigue mechanism tends to be greater than the sum of the degradation that would be caused by corrosion and fatigue acting separately.
- It rarely occurs at a single location; once corrosion-fatigue is found it is highly likely that it has already occurred at other locations.

Both these duplex mechanisms can be responsible for widespread degradation of boiler components. They can therefore be responsible for 'life-limitation' of any of the main components – but the fact remains that they are still highly unpredictable mechanisms; sometimes they will occur, sometimes they will not.

14.5 Life assessment methods

We have seen that there are several possible failure mechanisms that can act as 'life-limiters' for boiler components. The situation is further complicated by the revelation that these mechanisms rarely act singly – failure is nearly always the result of some combination of events (the duplex mechanism). This potentially makes the assessment of component lifetime difficult; how can we find a rational answer within such a complex picture? The technique lies mainly with *predictability*. The mechanisms of creep and fatigue have a certain predictability – this means they can be analysed using quantitative methods, whereas corrosion, in its multiple forms, does not and must therefore be treated in a more qualitative way. These quantitative methods, although simple, have come to be accepted as an important aspect of both initial design life and remnant life assessment techniques.

We must now take an important step. Put to one side the notion that there is one single, somehow invariably accurate, way of predicting component lifetime. There is no *correct* way, but there are various *different* ways of approaching the realities of material behaviour. Try not to confuse this with a set of conventions: they are not conventions, as such; they are *approaches*. One of the key precepts of these approaches is the fundamental assumption as to whether a material is assessed as if it has 'pre-existing defects', or whether it has a truly homogeneous and unflawed structure. A further one is the assumption of the way that the material behaves at the tip of the crack – the two major alternative approaches here are (see Fig. 14.4):

- The material behaves in an absolutely elastic way. This encourages the use of the so-called linear elastic fracture mechanics (LEFM) approach with its related concept of the fracture toughness (K_{1c}) parameter.

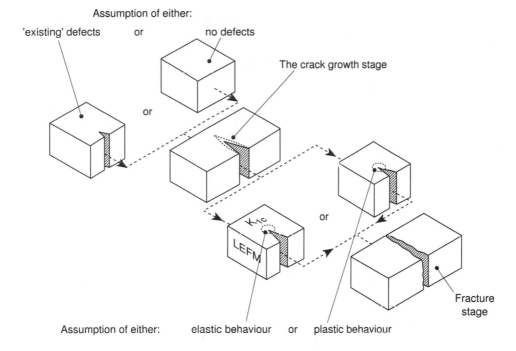

Points of agreement (in all approaches) are:
  • crack *size* is important: large cracks cause quick failure;
  • cracks always grow: they do not get smaller;
  • all damage is *cumulative*.

Figure 14.4 How metals fail – different approaches

- It behaves in a totally plastic way at a crack tip, leading to a rather complex set of situations referred to generically as 'plastic collapse'.

There are fundamental differences between these two approaches, both of which have a sound theoretical basis. In order not to complicate the picture further it is best to take a small conceptual step backwards and look at those assumptions and precepts which are *common* to the various approaches to material behaviour. Fortunately, there are four of these:

- The importance of *crack size*. It is generally accepted that when a crack in a component reaches a certain size, it will cause the component to fail under its normal working stress. This is known, rather loosely, as the *critical crack size*.
- Damage is *progressive*. Behind this rather obvious statement lies the fact that material defects get progressively larger as a material progresses through its lifetime towards failure. Cracks do not get smaller, go away, or repair themselves.
- Robust data is difficult to accumulate. Data such as creep rupture strength relative to temperature and time exposure has a high degree of scatter. It is notoriously difficult to correlate from results of similar tests and reproducibility is poor. This is an area where technical data can only be taken as *indicative*.

- Damage due to degradation mechanisms, whatever they are, is always *cumulative* – their effects add together.

These common factors form a useful 'starting point' for practical techniques of life assessment. International boiler standards use them as the basis for a simplified and commonly used approach. This is based on calculation of creep and fatigue 'life fractions', incorporating conservative margins to allow for the technical uncertainties explained previously. Further factors and margins may then be used to allow for the less predictable effects of corrosion-based mechanisms. Although such methods can be applied to all pressure-parts of a boiler they give better results for thick-walled components such as headers, rather than tubes. There are two parts to the methodology which are described below.

Calculating creep life

The key criterion here is creep rupture strength. Published data sources show how creep rupture strength reduces significantly with exposure to high temperatures. A typical curve for a superheater alloy steel is shown in Fig. 14.5. Note how the horizontal axis extends from zero to 250 000 hours. In most boiler standards it is normal to apply a factor of 0.8 to the *lower* bound of the data scatter-band to allow for errors.

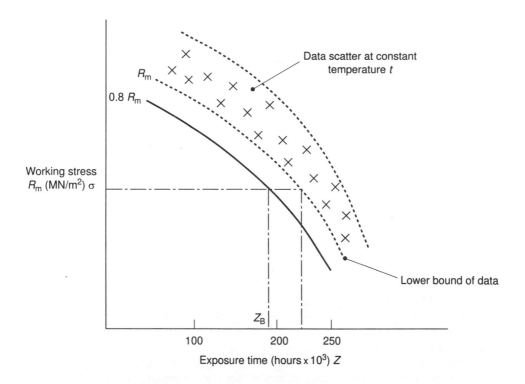

Figure 14.5 How to estimate creep life

Hence, referring to Fig. 14.5,

Lower bound is $R_m(Z, t)$ where Z = exposure time
t = mean wall temperature
Curve used is $0.8R_m(Z, t)$.

From the curve, the maximum design life corresponding to a particular working stress level can be read off.

The symbol normally used for creep life fraction is e_z. This is calculated by summing the cumulative exposures at various 'mean wall' operating temperatures that the component has seen. You will see this expressed as:

$$e_z = \sum_k e_{z,k} \quad \text{where} \quad e_{z,k} = \frac{Z_t}{Z_{Bt}} \times 100\%$$

All this means is that the individual lifetime fractions used up at different temperatures are added linearly together to calculate the overall cumulative effect on the lifetime.

Calculating fatigue life

Fatigue consists of two elements: that produced by a cyclic mechanical stress range and that resulting from thermal stresses caused by temperature differences. Fatigue life fraction is represented by e_w. As with creep, fatigue life data is normally taken from published sources. Fractions are again added linearly as:

$$e_w = \sum_k e_{w,k} \quad \text{where} \quad e_{w,k} = \frac{n_k}{n} \times 100\%$$

This infers that it is necessary to add the fatigue effects of stress cycles to those of thermal cycling to achieve the cumulative fatigue result.

Calculating total lifetime

In most practical lifetime assessments it is assumed that fatigue damage and creep damage act cumulatively to 'use up' the safe working lifetime of a component. There is some empirical evidence for a synergy effect but this is difficult to quantify and best allowed for by using an overall design margin (remember that margins and factors are an important part of life assessment).

Hence, we can say that the total life fraction (e) is given by:

$$e = (e_z + e_w) \times F$$

where F is a factor applied to allow for unpredictable mechanisms such as corrosion synergy effects and technical uncertainties.

The value of F varies between technical boiler standards and between industries. Standard coal-fired plant typically uses $F = 1.4$, but this can be reduced progressively if the components are subject to non-destructive and metallurgical tests to *confirm* the material's condition. Safety-critical installations such as nuclear plant may use a value of $F = 2$ or 3 to give an ultra-conservative approach.

14.6 Case study tasks

Task 1: remnant life assessment

Figure 14.6 shows outline operating data for a large coal-fired boiler as described in the case study text. The boiler has been in use for approximately 20 years. From the data provided, and the background information provided in the case study text, carry out a quantitative remnant life assessment of the platen superheater headers. The objective is to use a conservative approach to allow for various data and information that is not available to you. You should present your results in tabular form so they could be easily understood by technical managers who may not be specialists in boiler design or metallurgy.

Task 2: stress equations

Given that the temperature distribution (T) through the header wall at radius (r) can be matched, quite closely, to the form $T = a + b \ln r$ where a and b are constants, develop the following general formulae that can be used to express radial and circumferential stresses caused by this temperature distribution:

$$\text{Radial stress } \sigma_r = A - \frac{B}{r^2} - \frac{\alpha ET}{2(1-v)}$$

$$\text{Circumferential stress } \sigma_H = A + \frac{B}{r^2} - \frac{\alpha ET}{2(1-v)} - \frac{E\alpha b}{2(1-v)}$$

These are equations that form an inherent part of international technical standards used to design boiler and pressure vessels (such as British Standard BS 5500 and American Society of Mechanical Engineers (ASME) design codes).

Methodology

Don't just read the case study text and then start the case study tasks. First ask yourself a question: 'Exactly what *kind* of problem is this?'.

True, there isn't much given information – and what there is, is rather thin. There is a quantitative aspect to it in the form of creep and fatigue life calculations, but these are relatively straightforward, although they are the core of the analysis. The character of the problem is defined by the existence of a large number of *technical issues* which, although not part of the case study tasks that you are asked to do, clearly have a significant effect on the safe remnant life of an engineering component. These are mentioned in the technical descriptions in the text. The more obvious ones are:

- Corrosion: this is a relevant life-limiting failure mechanism, but very unpredictable.

Boiler operating data

Total recorded operating hours since new:	74 290 hrs
Total number of hours on base-load duties:	41 610 hrs
Total number of hours on 4-hour two-shifting duties:	32 680 hrs

Platen superheater header data

Nominal design life: 190 000 hrs
Material: 13Cr-Mo4 alloy
Maximum design temperature = 500°C
Maximum measured metal temperature during base-load = 480°C
Maximum measured metal temperature during two-shifting = 520°C
Maximum design yield stress $(R_e) = 610$ MN/m^2
Maximum stress due to pressure = 360 MN/m^2
Maximum stress due to temperature gradients = 108 MN/m^2
Nominal 'design stress' = 468 MN/m^2

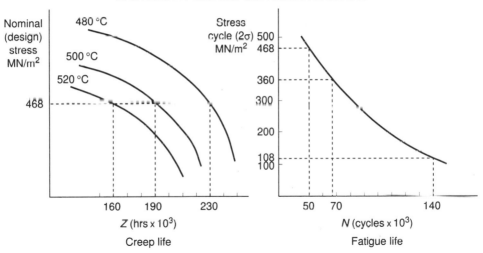

DATA FOR TYPICAL 13Cr-Mo4 HEADER MATERIAL

Figure 14.6 Boiler design and operating data

- Different approaches and assumptions about the way that materials degrade and fail. Life assessment demonstrates just how *un*prescriptive metallurgical theories can sometimes be.
- All the technical unknowns. No information is provided, for instance, about recent inspection results or previous problems with the boiler's components.

Given these observations, we can compare our problem of remnant life assessment (RLA) with the models presented in Chapter 2 and see if there is any match. RLA is certainly not a *linear technical* problem, as in Fig. 2.3; we have already decided that many of the surrounding technical issues have a high level of unknowns, and therefore cannot realistically be processed by a series of tight, quantitative, linear steps. Look also at the nature of the surrounding technical issues; complex though they may be, they are available to be addressed (take corrosion as an example – all you have to do is look for it). This means that we are not looking at the types of *nested* problem shown in Fig. 2.6(a) or 2.6(b). The model which best fits RLA problems is that of Fig. 2.6(c), the systemic type. The main clue is the existence of the surrounding technical 'satellite issues'. The best methodology to adopt is one which actively searches out these issues, which will only reveal themselves if you take active steps to expose them. This means that a pro-active approach is required, rather than a particularly careful or reflective one. Try to proceed on this basis.

14.7 Nomenclature

e_w	Fatigue life fraction
e_z	Creep life fraction
F	Factor applied to allow for uncertainties
n	Number of stress cycles for crack initiation
n_k	Number of load cycles during an evaluated period
$R_m(z, t)$	Rupture strength after time (z) at mean wall temperature (t)
Z_{Bt}	Design life at mean temperature (t)
Z_t	Exposure time at mean temperature (t)

15 The 'Schloss Adler' railway – design safety

Keywords

Safety design and embodiment design (the principles) – railway haulage system details – three safety principles: avoidance, protection and warnings. The fail-safe principle – the safe-life principle – design redundancy – design diversity – factors of safety. Embodiment design – defining 'systems' – safe embodiment design (five steps). Making design decisions.

15.1 Objectives – safety thinking

Figure 15.1 shows three pieces of engineering equipment. What is their purpose?

Did you make the mistake of assuming that the purpose of Fig. 15.1(a) is to act as a gear train, to transmit motion between shafts; why is it not also a finger-catcher? Figure 15.1(b) is undoubtedly a hydraulic buffer to absorb the kinetic energy of a moving rail truck – but could you design a better crusher? Figure 15.1(c) is obvious; it is a seat, or a picnic table, or somewhere to hide.

This chapter is about safety – more specifically safety in *design*. It is a commonly expressed belief that safety should be a consideration at all stages of the design process, from the earliest conceptual steps through to the final detailed engineering design. At the early stages, however, it is normal for safety requirements to be quite general – they only really start to take effect when things progress to the layout ('embodiment') and detail design stages. The case study introduces a set of *principles* of safety in design, then looks at how these are matched to the embodiment activity. It is good practice to concentrate on this embodiment step because this is where the rules and constraints that will influence the final detailed design of a component or system are set. We will be concerned, therefore, with the practical design aspects that help make a design 'safe', or at least reduce the risks as much as possible.

Safe design is also about *necessity*. Whatever the other pressures, a designer has little option but to produce designs which are safe, and can be shown to be so. The most aesthetic, optimised or innovative design will not last long if it is unsafe. Unfortunately, the activity of *designing for safety* is a hybrid one, involving not only well-proven safety practices (and there are many) but also an element of anticipation – you need to look forward to ask what the safety risks might be. They will not

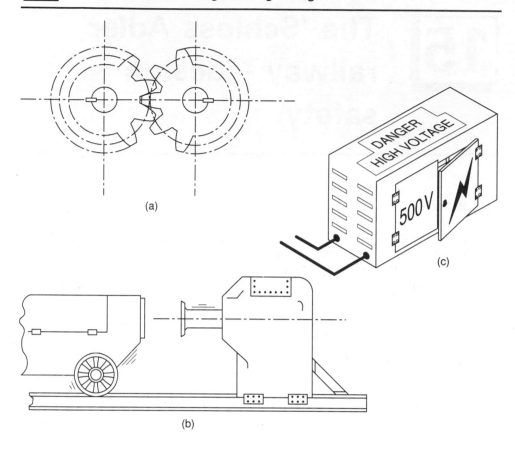

Figure 15.1 Engineering features – or dangers?

always be obvious. For this reason, design safety is a family of *closed problems*, a special type introduced in Chapter 2 of this book. The 'closed problem' methodology is one of the keys to obtaining good solutions to design safety problems – the challenge is to use it in combination with the general engineering-based principles of embodiment design. This makes for an effective approach.

15.2 The problem

The problem with the 'Schloss Adler' is access (look at Fig. 15.2). Situated on a basalt crag protruding 200 metres from the surrounding countryside it is inaccessible by motor vehicle. There are simply no roads to it. Part of the attraction to tourists is that visitors are transported by cable car from the station, situated near a small picnic and recreation area at the foot of the crag. Stores, fuel (coal and logs) and general building materials for the ongoing maintenance of the castle are transported by a single track railway haulage system. This was cut through the solid rock, following natural fissures, when the castle was built and originally used horses for motive power. The current electric-hauled system is 80 years old, unreliable, and badly in need of upgrading.

15.3 The railway – technical details

Figure 15.2 shows the general layout of the haulage system. The tunnel was designed to accommodate a single 800 mm narrow gauge track and a rail truck containing a maximum load of 1000 kg, including oversized items such as lengths of wood and building materials. There are three stops ('stations'): the top one in the castle's kitchens, an intermediate one at the fuel bunker and the lower station next to the cable-car station. The two upper stations are close to each other; a track length of only 5 metres separates the stopping points of the rail truck. A long, unimpeded run of approximately 250 metres leads down to the lower station. There is a short track spur at the lower station, used to store the spare rail truck. The track junction has manually operated 'points' and the tracks have buffers at their ends.

Braking

The truck is hauled up the track by an electric winch and braided steel rope. After manual unloading, the truck returns by gravity, pulling the rope from the winch as it goes. Some natural braking is provided by the friction of the winch and pulley mechanism but the main braking force comes from an electrically operated friction-drum brake installed on the winch shaft. This is triggered by mechanical trip switches situated several metres above the intermediate and lower stations.

Control

The control system has to be operated by unskilled labourers and kitchen staff and be simple (see Fig. 15.3). The truck is manually loaded at the lower station – the operator can then despatch it to either the fuel station or the kitchen, where it is unloaded by another operator who dispatches it back down again. Either of the upper stations can be supplied like this. A signalling system is needed to enable an operator at any of the stations to know where the truck is – there is a telephone system linking the stations but staff are still always complaining that they don't know where the truck is, and whether it is full or empty. It is not unknown for a truck which is still half-full to suddenly disappear from the upper stations, having been summoned by the operator at the bottom station, keen to send the next load. The electrical supply for the control and signalling system uses the castle's domestic 240 V a.c., a converter having recently been installed to produce a more powerful three-phase supply for the winch.

Some past problems

This has not proved to be the world's safest railway system. There have been several derailments, both in the tunnel, caused by excessively long loads fouling the tunnel sides, and at the bottom station, caused mainly by 'runaways' after the rope has slipped off the hook attaching it to the truck. After persistent runaways, heavy steel

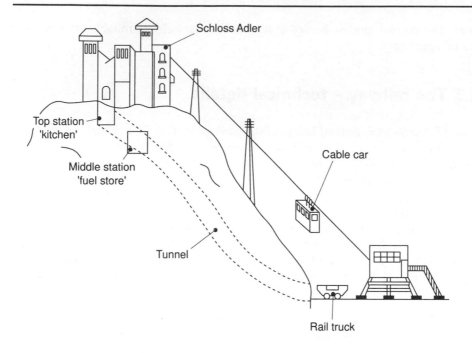

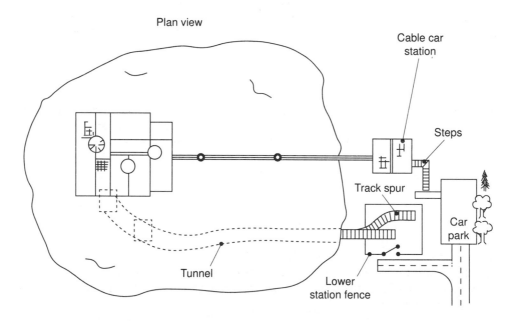

Figure 15.2 Railway system – general layout

buffers were welded to the lower end of the track run, to stop the loaded truck hurtling off the end of the track and disappearing into the forest (via the car park), accompanied by showers of potatoes and bags of cement. Luckily no serious injuries

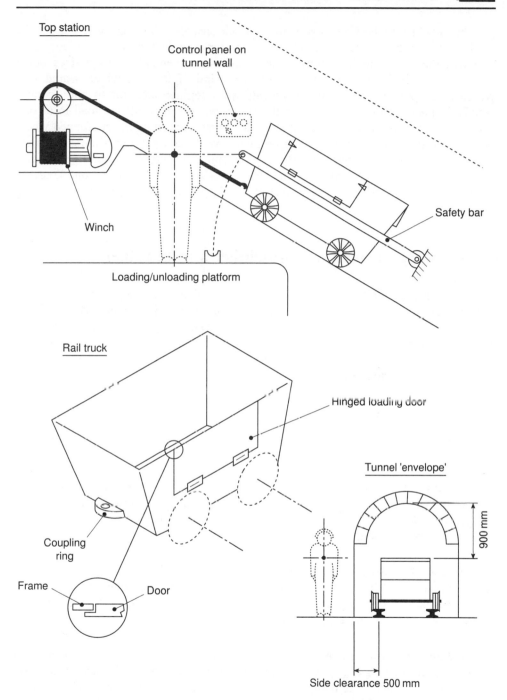

Figure 15.3 Further layout details

had been recorded and such incidents became known jokingly amongst the staff as 'runawaydays'. Legends were built up as to how far the truck had travelled into the forest before coming to rest. The buffers had stopped all that though.

There could also be problems at the top stations; the main one was getting the truck to stop in the correct place. The winch motor often had to be inched repeatedly to move the truck the last metre or so – this was eventually accepted as being governed largely by fate, the truck having a mind of its own which would stop wherever it wanted. Minor misadjustments were compensated for by the operator simply leaning out precariously over the single guard rail to lift the load in or out. Thursdays, for some reason, seemed to be the worst – the truck never stopped where you wanted it to on Thursdays. Other problems were mainly minor affairs: electrical short-circuits caused by water on the exposed junction-boxes, blown signal lamps and the odd tourist's dog straying into the tunnel from the lower station. Tourists' dogs and children were always a frustration; they just *wouldn't* read the warning signs.

15.4 Safety in design – principles and practice

Straight thinking about safety in design involves linking together the principles and the practice, all underpinned by the action of *anticipation* introduced earlier. These principles and practices all have an engineering basis tempered by common sense. We will look at these in turn, using some straightforward examples – remember that they are related more to the embodiment phase of a design, rather than earlier conceptual design activities.

Safety principles

There are three fundamental principles of design safety, as shown in Fig. 15.4. These are, in *ascending* order of effectiveness: warnings, protection and avoidance.

Warnings   Warnings are the least effective method of trying to ensure design safety, but still the most common. Whilst signs and instructions may warn of a danger, they don't make it go away, or make its effect any less. Warnings are passive and don't work particularly well.

Protective features   This is an indirect method of design safety. It comes in many forms but all have the characteristics of relying on *protecting* people from a danger (or failure) or mitigating its effects in some way. The danger itself, however, does not go away. Some typical examples are:

Control and regulation systems   These give a basic form of protection by keeping the operation of a process or item of equipment within a set of pre-determined safe limits. This minimises the danger of failure. An example is the combustion control system on a boiler which acts as a safety system because it limits, albeit indirectly, the steam pressure.

Design diversity   This is a more subtle principle. The idea of using different design principles within the same design gives protection against common-mode faults and failures. It almost guarantees that any fault that does occur unexpectedly will not be

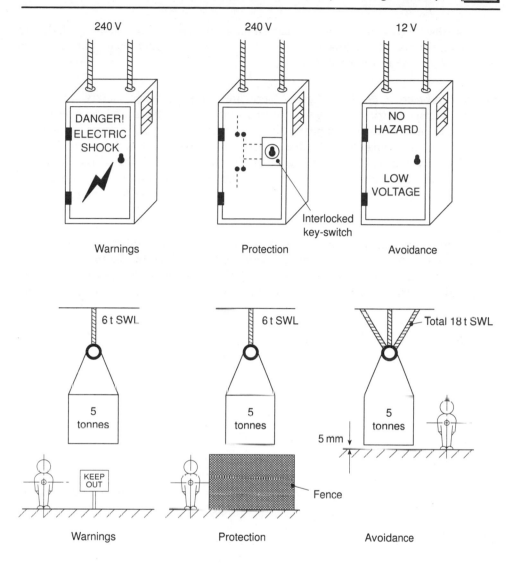

Figure 15.4 Principles of design safety

replicated throughout the entire design – so it will not 'fail' completely. Computer software systems are a classic example of this approach but it is also employed in mechanical equipment by using different materials, types of bearings and sealing arrangements, fluid flowpaths, speeds, etc.

Factors of safety   Factors of safety, design margins, conservatism, prudence – these are all protective instruments, used to reduce the risk of dangers caused by failure. They apply to mechanical, electrical and electronic systems. Sometimes they are carefully calculated using known properties and failure modes but at other times they are chosen as a *substitute* for detailed design knowledge. In all too many engineering designs, data on material performance or the state of loading of individual components, and a unanimous understanding of how things actually fail, are imperfect. Fatigue life calculations are a good example in which factors of safety are more *palliative* than prescriptive.

General protective devices Again, there are different types, safety valves on boilers, overspeed trips on engines, overcurrent devices, fuses and reverse power protection on electrical equipment being typical examples. Larger equipment installations have separate protective systems, often with features such as duplication, diversity and a self-monitoring capability, to keep the level of risk down. A related type of indirect design safety feature is the *lock-out*; this prevents a component or piece of equipment which is in a dangerous state from being put into operation.

Avoidance features This is the principle of achieving safety by choosing a design solution that eliminates danger from the outset. It is by far the most direct (and best) method of ensuring design safety — although not always possible. Eliminating potential dangers starts at the embodiment stage of the engineering design process and feeds forward into the engineering specifications for a piece of equipment. There is often a link, of sorts, with some of the protective features described previously. You can think of danger avoidance features as fitting neatly into three separate principles: fail-safe, safe-life and redundancy.

The fail-safe principle This is not quite the same as having protective devices. The idea is that a piece of equipment is designed to *allow for* a failure during its service life but the design is such that the failure has no grave effects. The failure is *controlled*. There are a few ways to do this:

• First and foremost, there must be some way of identifying that the failure has happened — it must be signalled.
• The failure must be *restricted*, i.e. for a machine, it must keep operating, albeit in a limited or restricted way, until it can be safely taken out of operation without causing danger.
• The implications of failure of a single component need to be understood, assessable, as to the effect that it will have on the total machine or system design.

All these presuppose that the consequences of failure are properly understood by the designer — a precise understanding of the definitions is less important than the need to develop a clear view of how a component can fail and what the consequences will be. A useful tool to help with this is the technique of *fault tree analysis* (FTA). You may see variations of this; a common one is failure modes effect analysis (FMEA). There are small differences but the principles are the same. To perform an FTA, you list all the possible modes of failure of a design, and display the consequences of each failure in a network or 'tree' diagram. Figure 15.5 shows an example for a bolted steam pipe joint. Note how the tree starts from the smallest, most divisible components and moves 'outwards' to encompass the design of the 'system' (the bolted joint). There is a useful methodological principle in Fig. 15.5. Look at how the joint is considered as a *system* with joined and interlinked parts. Try to connect this with Fig. 2.6 in Chapter 2 — this type of systems thinking is Fig. 2.6(c) in action. If

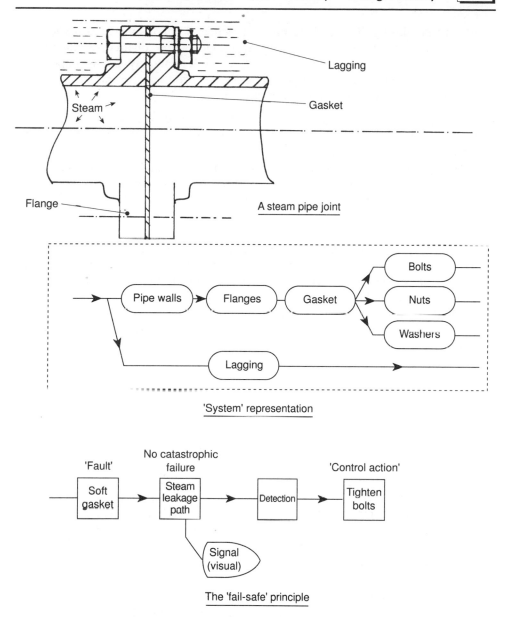

A steam pipe joint

'System' representation

The 'fail-safe' principle

Figure 15.5 A typical component system analysis

you can learn to think like this, you will find it easier to identify potential problems in design safety, and eliminate them at an early stage. FTA is of greatest use on complex designs consisting of interlinked and nested systems; oil and gas installations, nuclear plant, airliners, weapons systems and most electronic products are assessed using FTA techniques.

The safe-life principle This sounds rather obvious: all design components (mechanical, electrical or electronic) need to be designed for adequate *lifetime*, so they won't fail during their working life. In practice, it is not so easy as it sounds — safety is based exclusively on accurate quantitative and qualitative knowledge (yours) of all the influences at work on a component. It is, frankly, almost impossible to do this from scratch every time you do a design — you have to rely on previous, proven practice. There are four areas to consider:

- Safe *embodiment* design based on proven principles and calculations. This means technical standards and codes of practice — but note that not all of them contain embodiment design details.
- Careful specification of *operating conditions*. Operating conditions for engineering components have to be described fully. Fatigue, creep and corrosive conditions are important for mechanical components as they have a significant effect on material lifetime. For electrical equipment, environmental conditions (heat, dust, dampness, sunlight, etc.) can soon reduce lifetime.
- Safe operating *limits* — again, it is easy to overlook some of the operating limits of engineering components. Low-cycle fatigue and high-temperature creep cause the most mechanical failures if they are overlooked at the design stage. Stresses due to dynamic and shock loadings are another problem area. These failures often occur well within the estimated lifetime of a component.
- Analysis of *overload conditions*. It is not good enough just to consider normal working stresses, currents or speeds; you need to look for the overload condition.

Many mechanical equipment designs have an accepted way of calculating their projected lifetime. Contact bearings are designed using well-proven lifetime projections expressed as an 'L-number'. Technical standards for high-pressure boilers and steam vessels specify calculation methods for creep and fatigue life. Safety-critical items such as structures for nuclear reactors, aircraft, tall buildings and high-integrity rotating plant are also designed in this way. There are, however, numerous items of equipment for which the technical standards do not address lifetime at all. The common mechanical engineering standards covering, for example, steel castings and forgings place great emphasis on specifying detailed mechanical and chemical properties but hardly mention fatigue life. To compensate, manufacturers of specialised forged and cast components do in-house tests and develop their own rules and practices for defining (and improving) component lifetime.

The principle of redundancy Redundancy is a common way of improving both the safety and reliability of a design — it is also easily misunderstood. The most common misconception is that incorporating redundancy always increases design safety but there are many cases where this is not true. What do you think of the following statement?

- An airliner with four engines is safer than one with two engines.

There appears to be some logic in this. On long-haul Atlantic routes the theory is that a four-engined aircraft can suffer two engine failures and still complete the journey safely using the two remaining engines. This is fine for some types of engine failure, but what if an engine suffers a major blade breakage and the damaged pieces smash through the engine casing into the wing fuel tanks? Here the redundancy has no positive effect, in fact there is a counter-argument that four engines have twice the chance of going wrong than do two. The central message is that redundancy is not a substitute for the proper design use of the fail-safe and safe-life principles. Redundancy *can* increase design safety, but cnly if the redundant components are themselves designed using fail-safe and safe-life considerations. Some specific examples of design redundancy are shown in Fig. 15.6. Note the different types, and the definitions that go with them. These definitions are not rigid, or unique – their main purpose is to help you think about and identify the different options that are available.

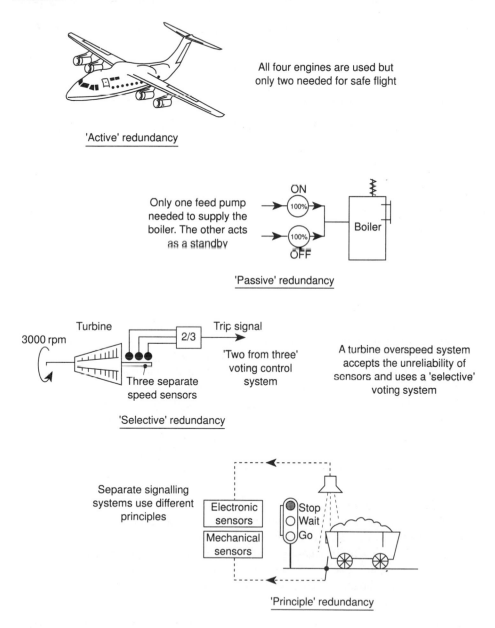

Figure 15.6 Types of design redundancy

The danger	Comment
Stored energy	Potential and kinetic energy can be dangerous. Stored pressure energy is a particular hazard if released in an uncontrolled way.
Rotating machinery	Rotating belts, couplings, gears, fans – anywhere where there is relative movement between a machine and humans is a potential danger.
'Crushing and trap' gaps	Gaps of more than about 8 mm between moving parts can crush or trap fingers.
Exposed electrics	Exposed electrical equipment is an obvious danger.
Hot parts	Components or fluids above 50°C will cause burns.
Falling and slipping places	If there is a place where it is possible to fall or slip, someone will always find it.
Noise	Excessive noise is a recognised industrial hazard.

Figure 15.7 Typical design dangers

Embodiment design safety

Embodiment – a rethink Embodiment design is what happens in the rather large grey area between conceptual design, and detailed engineering design where a system or machine is described by a set of specifications and drawings. Embodiment is therefore about *deciding engineering features*. Hence deciding safety features is part of the embodiment process but is not all of it; other engineering considerations (the main one being *function*) have to be included as well. The process of identifying general design safety features is easier than choosing between all the available alternatives. This is because, unlike, for example, the mechanical strength of a component design, safety cannot easily be expressed in quantitative terms, so you often have to work without clear-cut acceptance criteria that you can use to compare the safety level of different designs. It is easy to make general statements about design safety, but not so easy to translate these into the language and features of embodiment design. The best place to start is with a list of design *dangers* and then consider them as you think, in turn, about each part of a design. These design dangers are different, in detail, for each individual component or system design, depending upon what it does, but the general principles are common. Figure 15.7 shows a typical list of design dangers. We can use these as part of a series of steps to tease out good safety features during the embodiment process.

***Step 1: split the design into* systems** Any technical design, simple or complex, can be thought of as consisting of interconnected systems. These systems come together to make a design 'work'. The methodology of this should be easier to understand by studying Fig. 15.8, a simple passenger lift showing the design broken down using this type of systems approach. The lower part of the figure shows the function of the lift split into three primary systems: structure, mechanics and electrics. This primary system allocation is the most important, so note two points:

- Are you comfortable with the way that the lift shaft and car *structure* is shown as a system?

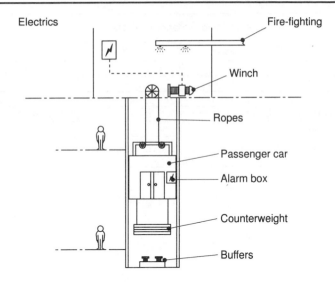

Electrics

Fire-fighting

Winch

Ropes

Passenger car

Alarm box

Counterweight

Buffers

PASSENGER LIFT

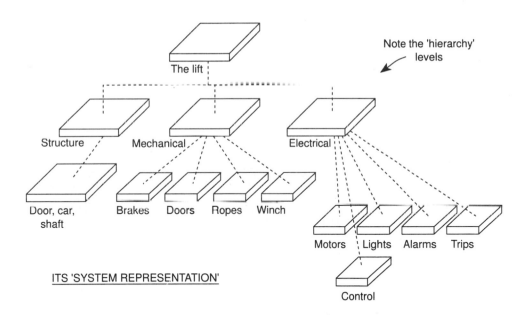

The lift

Note the 'hierarchy' levels

Structure   Mechanical   Electrical

Door, car, shaft   Brakes   Doors   Ropes   Winch

Motors   Lights   Alarms   Trips

Control

ITS 'SYSTEM REPRESENTATION'

Figure 15.8 Representing a design as 'systems'

- A 'system' does not have to be a process, or an electrical or control network. Structures and mechanical components can also be thought of as systems.

Each primary system is then subdivided into its constituent subsystems; this gives a better resolution of what each subsystem actually does. In theory, you could go on indefinitely subdividing systems down the levels – a good practical approach. For the purposes of looking at embodiment design, it is best to use no more than three levels; further subdivisions will over-complicate the analysis.

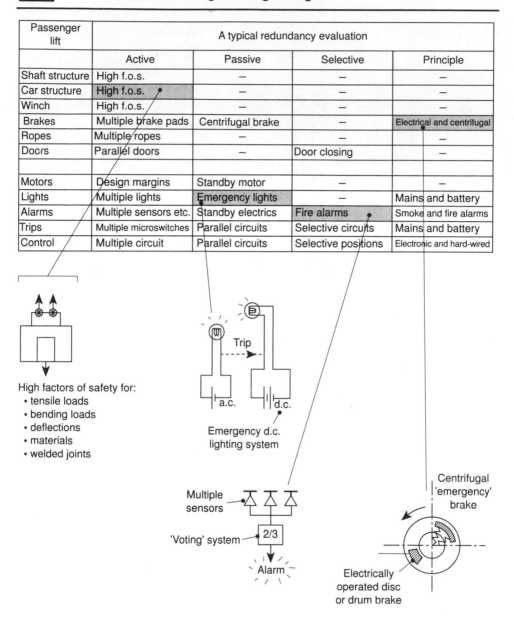

Passenger lift	A typical redundancy evaluation			
	Active	Passive	Selective	Principle
Shaft structure	High f.o.s.	–	–	–
Car structure	High f.o.s.	–	–	–
Winch	High f.o.s.	–	–	–
Brakes	Multiple brake pads	Centrifugal brake	–	Electrical and centrifugal
Ropes	Multiple ropes	–	–	–
Doors	Parallel doors	–	Door closing	–
Motors	Design margins	Standby motor	–	–
Lights	Multiple lights	Emergency lights	–	Mains and battery
Alarms	Multiple sensors etc.	Standby electrics	Fire alarms	Smoke and fire alarms
Trips	Multiple microswitches	Parallel circuits	Selective circuits	Mains and battery
Control	Multiple circuit	Parallel circuits	Selective positions	Electronic and hard-wired

High factors of safety for:
• tensile loads
• bending loads
• deflections
• materials
• welded joints

Trip

Emergency d.c.
lighting system

a.c.   d.c.

Multiple sensors

'Voting' system — 2/3

Alarm

Centrifugal
'emergency'
brake

Electrically
operated disc
or drum brake

Figure 15.9 A redundancy evaluation

Step 2: consider redundancy Try to think about redundancy before you get too involved with the details of individual parts of a design. The best technique is to list the various options, showing how they can apply to each system. Figure 15.9 shows a sample for the passenger lift design – note how it incorporates each of the different types of redundancy introduced in Fig. 15.6.

Step 3: list the danger features You have to do this system by system. Any attempt to short-circuit the exercise by trying to 'home in' intuitively on the danger features that you feel are most obvious, or think of first, will not give the best results. Its steps need to be a systematic exercise – this is the whole key to opening up the problem (remember Fig. 2.5 in Chapter 2?) and being able to identify *all* the design safety features, not just the easy ones. Figure 15.10 shows a typical analysis of danger features for one of the systems of the passenger lift. First you have to identify a danger feature, then do something about it. Now is the time to introduce the principles of design safety discussed earlier: avoidance features, protective features and warnings, in this order of preference. Good, objective embodiment design is about eliminating design safety problems at source, before they get *into* the design. It is a proven fact that once a feature becomes an accepted part of a design, and progress is made through the detailed engineering stage, it becomes very difficult to change it, without having to make changes to other (probably desirable) design features also. Natural reaction is to leave the original design features in; hence the only way to cover up the dangerous design feature is by *protection* – a less than ideal method, as we have seen. The message is simple: danger features must be designed *out* at the embodiment stage, rather than trying to cover them up later. You can infer the basic methodology by a close look at Fig. 15.10. Although there are no rigid rules, you can generally get the best results by looking at the features in the order shown (i.e. working left to right across the table). Note that the figure only shows the analysis for one of the systems – for a full embodiment analysis, the process would be repeated for all the other systems identified previously.

Step 4: look for embodiment options   It is rare that the first embodiment design ideas will be the best ones. The principle of finding the best solution lies with the activity of 'opening up' the design – revealing its complexity – to find the most appropriate solution from the 'possibles' available. Although the embodiment design activity benefits from a certain level of innovative thinking it is important not to confuse this with true innovation. True innovation belongs in the conceptual stage that *precedes* the embodiment design steps – embodiment is a more prescriptive, better-defined process than this. Innovative embodiment solutions are fine, as long as they fit within the constraints of using those conceptual design decisions that have already been made. Here are two examples to do with the passenger lift in Fig. 15.8, to reinforce this point:

- The conceptual design is for a 'rope operated' lift, not an alternative design operated by hydraulic rams. Hence the embodiment design should accept the use of an electric winch and look for ways to make the electrical system safe. It would be wrong to suggest the use of a hydraulic lifting system instead. That would be interfering with the agreed concept design.
- If, for example, the passenger lift was designed with double sets of doors, the embodiment design stage should accept this fact and look for options to make the double-door arrangement safer, not ways to change the concept to one using only a single set of doors.

This is a practical constraint to help avoid the design process descending into a state of anarchy, not a mechanism to discourage innovation.

Figure 15.11 shows typical embodiment design options for the passenger lift, looking at the safety problem of crush/trap gaps between and around the lift's inner sliding doors. Note how sketches are used rather than detailed written descriptions.

PASSENGER LIFT DOOR SYSTEM — EMBODIMENT/SAFETY DESIGN FEATURES

| Danger/feature | | Avoidance | | Control | Protection | | Warnings |
		Fail-safe	Safe-life		F.O.S.	Protection system	
Stored energy	Door closing force	Soft door-seals	Slow closing	Manual opening facility	Low closing force	Proximity interlock for doors	Yes
Rotating parts	External pulleys	Pulleys outside the car	N/A	N/A	N/A	Maintenance interlock	No
Crush and trap gaps	Between doors: top and bottom gaps	< 5 mm gaps	N/A	Manual opening	N/A	Proximity interlock for doors	Yes
Exposed electrics	Behind car panel only	Low voltage only	Fuses	Control all automatic	N/A	Interlocked electrical doors	Yes
Hot parts		N/A	N/A	N/A	N/A	N/A	No
Falling and slipping	Slips inside car only	No access to shaft	N/A	N/A	N/A	Non-slip flooring and hand rail	No
Noise	Low noise levels only	Remote winch	N/A	N/A	N/A	N/A	No

N/A - not applicable
Safe-life criterion comprises:
– proven principles and calculations
– assessment of operating limits
– analysis of overload conditions

Figure 15.10 Embodiment design – danger features

There are two reasons for this: first, sketches help *definition*; they can capture and define that fleeting idea in a way that description cannot. Secondly, sketches are a better way to *communicate* embodiment design ideas to other people. Other viewpoints are often necessary to help 'firm up' ideas. It is worth using sketches.

- Embodiment design is about specific features, not general principles – so use *sketches*.

PASSENGER LIFT–DOOR system: EMBODIMENT DESIGN				
Danger	Embodiment	Good points	Bad points	Alternatives?
Stored energy: Door closing force	10 → Electrical load-limiter ← 10 Max nip force = 10 N	Simple electrical limiter	Spurious signals (continual reopening of doors)	Variable closing force
Crush and trap gaps	Car body < 5 mm top and bottom gaps to eliminate finger traps Doors	Safe: no further guards required	Jamming. Accurate manufacture needed	Cover-plates Alternative guarding arrangement
Crush and trap gaps	Doors Beam Infra-red 'proximity' beam cuts off door movement Side view	No moving parts	Unreliable. Sensors easily damaged	Pressure-pads Physical barrier
Exposed electrics	Control/alarm panel inside lift car Low Voltage (LV) High Voltage (HV)	No HV danger. LV works well for these functions	Transformers required. More complex wiring	Use HV but include increased protection devices

Figure 15.11 Embodiment design options

Step 5: deciding It is no use defining lots of elaborate design options if you can't then decide which ones to use in the final detailed design. A gamut of terminology surrounds this step; you will see it referred to as 'design evaluation', or 'design synthesis', through to the more elaborate 'evaluating concept variants'. All mean the same: *deciding*. But how? Sadly, there are no hard and fast rules; it would be nice if there were. Several factors impinge upon the decisions; you can analyse and weight design options in order of safety, discuss them, eliminate the worst ones – but you will still need to apply some intuition and experience. There are a few broad guidelines to follow:

- *Avoid contradictions*: make sure the technical choices are consistent with each other. There would be little logic, for example, in designing one electrical circuit for low-voltage operation, for safety, if others nearby operated on high voltage. Avoid contradictions by aiming for consistency.

- *Use technical standards*: these are useful to help you decide. Design details in published technical standards have invariably been subject to long discussion between people at the sharp end of manufacture and use of the equipment in question. You can rely on standards therefore to provide *proven advice*. The main limitation is in the scope of technical standards – not all technical standards cover embodiment design, some being intended more as purchasing and specification guides rather than a design tool. This means that some technical standards are more useful than others.

- *Use technical guidelines*: these are available in many forms: data-books, nomograms and manufacturers' publications. The quality of such information varies widely; data-books can be particularly useful for embodiment design ideas. Manufacturers' catalogues are good at showing different embodiment designs that are available (itself an indication of the success of the design feature) but tend to be optimistic in under-estimating any negative aspects of a design option.

- *Consider the cost*: more specifically, cost *effectiveness*. Frankly, you have to develop an instinct for this. Cost is a real constraint in all design projects and whereas features such as redundancy and diversity are always desirable, it is not economic to duplicate everything. Your objective should be to keep a focus on cost-effectiveness when deciding embodiment design – but keep it in perspective.

These aspects of deciding embodiment design should be treated as guidelines only. They must be seen as lying within the overall objective of making embodiment design as safe as possible, but also simple. Simplicity is a desirable design characteristic; good, safe, designs are often very simple.

15.5 Case study task

The railway haulage system installed in the 'Schloss Adler' is inadequate by modern safety standards. It is a particularly difficult installation from a safety point of view, combining the inherent dangers of an industrial railway with the proximity of the public. This is not a good mixture. For practical and cost reasons the system can only be refurbished, rather than rerouted, so it must use the existing tunnels and station locations. The refurbishment is not intended to change the *conceptual* design, as outlined in the technical details section of the case study text. Many of the embodiment design details, however, are clearly capable of improvement. The case study task is to improve the *design safety* of the Schloss Adler rail haulage system. This involves applying the principles and practices of design safety, as explained in the study text, to the various aspects of embodiment design.

Methodology

A good methodology is to follow the five steps outlined in the study text:

- split the design into systems;
- consider redundancy;

- list the danger features;
- look for embodiment options;
- make some design *decisions*.

The most effective way to record information during these steps is by using *sketches*, similar to the examples shown in Fig. 15.11. Detailed commentary is not required. Sketches may be annotated to help explain a design feature, but keep it short. Because of the complexity of some of the systems in the railway installation, group working is a good way to approach this case study task.

16 Electric vehicles – design for plastics

Keywords

Background to car design – bodyshell and panel design – development pressures (fuel economy, recycleability and cost). Design with plastics (thermoplastics, thermosets, composites and alloys) – mechanical properties. Uses of plastics in cars – design criteria and common problems – the 'Citybee' electric vehicle.

16.1 Objectives

Advances in materials have an important effect on the design of engineering products. Materials technology is one of those areas of design that rarely remains static; small changes and improvements are being made all the time, accompanied by occasional larger development steps. Such steps are driven by the continual search for lighter, cheaper, stronger or longer-lasting materials for use in engineered products. Some of the biggest developments are made in the field of consumer products – an intensely competitive business sector where product innovation and advancement are essential, rather than a nice option, if a company is to remain in business.

This case study looks at one such example – the use of plastics in passenger cars. Cars present an interesting design challenge; they develop quickly, use complicated, multidisciplinary techniques and have to be a subtle blend of reliable engineering design and aesthetic appeal. They will only sell if they perform well and look good. The industry itself is very competitive, requiring vast capital investment to produce vehicles that, at best, are constrained to have a tight profit margin per unit sold. Because of the intense competition, car design moves quickly – one of the best examples being the use of plastic materials. Until the late 1970s most parts of a car were made of steel, only minor components such as trim strips and brightwork being made of other materials. If you look at any standard production-model car today you will find that many of the components are now made of plastic-based materials. These include not only trim components but also 'under-bonnet' parts (including some engine components) and structural body panels. These material changes have had a significant effect on the design of cars – they have affected function as well as looks.

16.2 Car design

The design of modern production cars starts with the aerodynamic requirements of the bodyshell and the space constraints of the engine and passenger compartments.

The bodyshell is of 'monocoque' design (i.e. it does not have a separate structural chassis). Some traditional car models retain the rear wheel drive arrangement, in which the shell floor has a recessed 'hump' to house the transmission and drive shaft. A more common arrangement is for a transverse mounted engine and front wheel drive – here the monocoque floor is flat. Bodyshells are made from deep-drawing low carbon steel with a tensile strength of about 300 MN/m^2 ar.d a Young's modulus of 200–300 GN/m^2. Thickness varies from 0.65 mm to 1.55 mm depending on the panel's location in the structure.

Aesthetic and aerodynamic considerations apart, the main function of a monocoque bodyshell is to provide the car with structural integrity. To do this it must be capable of distributing loads coming from the 'running gear' (wheels, axles and suspension) without too much deflection (bend or twist). Crash resistance is a key design feature – there are statutory requirements for this, so the shell must have enough rigidity to resist minor impacts but be designed to buckle in a serious crash. This absorbs the energy of the impact whilst the passengers are protected within a stronger, more rigid, 'cage'. All of these design requirements must be met without compromising the way in which the structure can be manufactured – to be viable, a bodyshell has to be suitable for mass-production methods.

The structural design of a bodyshell is done in two parts: overall shell analysis and individual panel conditions. Both use finite element (FE) methods and rely on various technical assumptions to derive approximations to structural behaviour.

Overall shell analysis

Figure 16.1 shows the principle of shell analysis. The bodyshell structure is considered as consisting entirely of members which act either as beams (which undergo pure bending) or elements which are in pure shear. The FE 'nodes' are defined at the intersection of the elements, each node then being subjected to the FE orthogonal equations during the computer analysis. The output of the analysis is a set of values for displacement of the node set. These displacements are substituted back into the equations to calculate the resulting stresses on each part of the structure. As with all FE techniques, accuracy depends on the quality of the input data that is fed in – correctly defined boundary loading conditions are particularly important.

Individual panel conditions

Individual panel stresses and strains (deflections) are calculated using well-known theories of thin-walled structures. Panels are assumed to fail by either compressive buckling or shear instability and deflections calculated accordingly. Although this is an idealised assumption, it forms a good approximation to reality, particularly for the shell panels at the rear of the car which are designed with heavy stiffeners around their edges to take the end-loads. Bodyshell design is not just about stresses and deflections – it has also to consider the way in which the shell can be manufactured. Panels are cold-stamped and then spot welded together to make up the monocoque shell. These processes are heavily automated. The manufacturing time for a bodyshell is on the critical path of the production schedule for all mass-produced cars – they are not bought in from subcontractors like many of the mechanical, electrical and trim components. Bodyshell manufacture is also expensive; tooling costs and

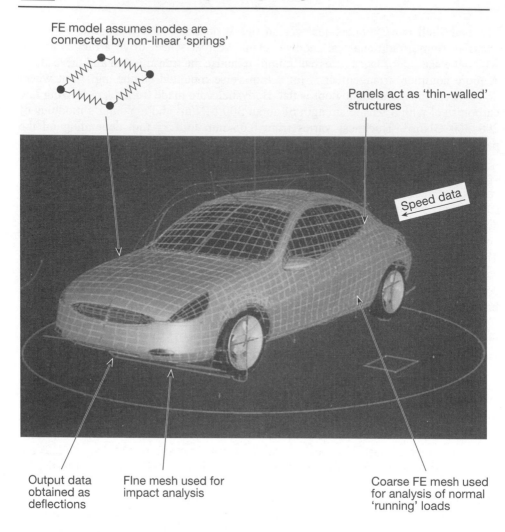

FE model assumes nodes are
connected by non-linear 'springs'

Panels act as 'thin-walled'
structures

Speed data

Output data
obtained as
deflections

FIne mesh used for
impact analysis

Coarse FE mesh used
for analysis of normal
'running' loads

Figure 16.1 Principles of FE body shell analysis

maintenance costs are high, because of the nature of the accurate stamping operations required.

16.3 The problem

Cars made from steel seem to look and work fine, so what is the problem? Does a problem really exist at all? You can find an answer by looking at the *business* pressures that affect car design. A car is a consumer product, rather than an industrial one, with the result that designs are predominately customer-driven. Coupled with the fact that there are too many car manufacturers compared to the market demand (there is worldwide overcapacity in manufacturing) manufacturers must both develop their design and strive continually to keep manufacturing costs down. These aspects are of prime concern and, taken together, form an important set of 'technical design'

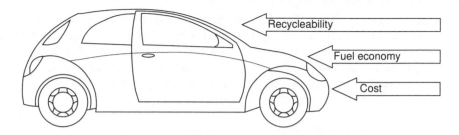

Figure 16.2 Pressures on car design

objectives for a car manufacturer. It is difficult to place them in an order or priority as all three contain a mixture of both technical and commercial implications. We will look at them in turn – whilst remembering that, in practice, they act together (see Fig. 16.2).

Fuel economy

There will always be pressures for cars to be more fuel-efficient because fuel is a finite, and therefore expensive, resource. Putting to one side developments in engine technology (this is an interesting, but different, subject), the easiest way to reduce a car's fuel consumption is to make it lighter. For an average four-passenger car, approximately 60 per cent of the engine power is needed just to move the car's own weight, without any passengers. Weight saving is therefore a very real technical objective. It does however carry constraints – weight reduction is only acceptable if it can be done without compromising the rigidity or crash resistance of the vehicle's bodyshell structure.

Recycleability

This is perhaps more of a technical *constraint* than an overt objective. Steel cars are almost perfectly recycleable and more than 90 per cent of the steel bodyshell panels, engines and other under-bonnet components are eventually recycled in some form. Other materials however (the main category is thermosetting plastics) are not recycleable, so there is always a balance to be achieved between the advantages and disadvantages of using new materials to replace steel.

Cost

Once again, we meet the spectre of cost and production realities of mass-produced engineering products. New designs and technical developments affect mainly three production cost categories:

- tooling costs
- production time costs
- materials costs.

The other major category, production labour costs, can be affected by design changes but normally only to a lesser extent.

Now we can recrystallise the problem. The challenge is how to improve car design whilst paying proper respect to the three main areas: fuel economy, recycleability (a constraint, remember) and production cost. Some of the answer lies with the use of plastics. Plastics are light and cheap and have some (but not all) of the strength properties of steel. They are easy to mould and shape, and some are recycleable. The difficulty lies in the choice of type of plastic (there are hundreds) and choosing which car components they are best suited for. There are many potential advantages, but also disadvantages if the wrong choices are made. The purpose of this case study is to address just this problem. It is predominantly a technical issue with the general characteristics of the type of 'linear technical problem' outlined in Chapter 2 (look at Fig. 2.3 if you need reminding of this common format of design problem). We will start by looking at some basic technical information about plastics and their use in cars.

16.4 Design with plastics

Every use of materials, however trivial or important, involves the activity of selection – so – ask your question:

- What is *so special* about designing with plastics?

First, some loosening up of ideas, and maybe a release from some misconceptions. Plastics are made up of many families of materials so it is misleading to think of them as a single group of materials differing only slightly in properties. Plastics designed for use in engineering applications, commonly called engineering plastics, are a technology world apart from the common types of cheap, low-strength 'commodity plastics' used in toys, household articles and low-quality household appliances. Engineering plastics have a wide range of very good technical properties including the important ones of strength, light weight and ease of processing. So:

- engineering plastics can be as useful as steel; and
- they are generally easier, more versatile, to work with.

This means that you can, on balance, produce better designs using plastics than you can with steel. You can be more innovative, think more conceptually and produce better design results.

It is worth taking a quick look ahead at what can be done with engineering plastics. One of their big advantages is that they can be injection-moulded to produce complex shapes – car dashboards are a good example. They can be reinforced with a fibre matrix to produce high-strength composite materials, as for body panels, and alloyed with other types of plastics to produce impact-resistant moulding for components such as bumpers. Practically, the difficulty of designing with plastics is the large number of alternatives that are available with similar-sounding names and abbreviations. It is important, therefore, to understand the main family groups of engineering plastics, and their properties and design applications.

There are many components in a car that can be made from plastics. Small moulded items such as switches, housings and connectors are obvious choices, as are trim items, door capping pieces, headlining, and the various types of clips and strips that make up the car interior. These are less important, however, than the areas in which plastics are used to replace steel in the design of load-bearing components. The main one is bodyshell panels. It is here that the cost and weight savings can be made, so we can treat them as being more important.

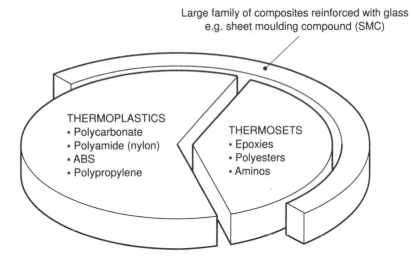

Large family of composites reinforced with glass
e.g. sheet moulding compound (SMC)

THERMOPLASTICS
• Polycarbonate
• Polyamide (nylon)
• ABS
• Polypropylene

THERMOSETS
• Epoxies
• Polyesters
• Aminos

Figure 16.3 Types of plastics

The terminology of plastics can be confusing. Many of the names are similar and it is, frankly, difficult to relate some of them to the materials themselves, and their most common applications. There is the added complication that many of the advanced alloyed plastics are referred to by manufacturers' trade names. In practice, the situation can be looked at rather more simply. There are basically only two families of plastic polymers: thermoplastics and thermosets. Thermoplastics are by far the most important to know about. There are also two further 'derived' categories: alloys and composites. The fundamental difference between thermoplastics and thermosets is the way in which they behave when heated – this gives them different properties and mouldability. We can look briefly at each in turn (see Fig. 16.3).

Thermoplastics

The key property of the thermoplastic family of polymers is that they can be resoftened by heating. This is because they have a loose structure, containing short molecules with weak intermolecular links. This softening means that processing the plastic is easier but has the disadvantage that the mechanical properties are affected by increased temperature – the material gets weaker. There are three main types of interest:

• Polycarbonate (PC): This is one of the most useful types of engineering thermoplastic; it is hard and resilient but its main feature is its toughness. It is used for moulded components that need to be impact-resistant. It is often alloyed with polyester.
• Polyamide (nylon): This is another common engineering material with good impact resistance and is often used for rotating wheels and sprockets. It benefits from being reinforced with a stronger material.
• Commodity thermoplastics: Two main commodity thermoplastics are acrylonitrile-butadiene-styrene (ABS) and polypropylene. As explained previously, these are not considered as 'engineering' thermoplastics because their strength values are too low.

There are three main moulding processes used for thermoplastics. The most common is injection moulding – you can think of this as the plastic equivalent of the pressure die-casting process used for metals – plastic granules are softened by heating and then injected into a mould. The moulding is completed to its finished state in a single operation and can include small features such as lugs, holes and threaded inserts. For three-dimensional components which are thin and hollow, the technique of blow moulding can be used – this works on the same principle as glass-blowing, the plastic being 'blown-up' inside a mould. The other process is compression moulding, in which the component shape is formed by compressing the softened material into a mould. It is useful for making near-flat bodyshell panels from reinforced polyester, a material known loosely as sheet moulding compound (SMC).

Thermosets

You will not meet these very often. The key property of a thermoset is that it cannot be resoftened by heating, owing to its strong cross-linked molecular structure. If it is reheated it will char or burn rather than melt. Traditionally, thermosets are considered difficult to use; only a few types can be processed by injection moulding and compression moulding.

Typical thermosets are:

- epoxies
- some polyesters
- the amino family, e.g. phenol formaldehyde.

Composites and alloys

Composites are basically a plastic matrix reinforced with a strong fibre material. The plastic matrix can be either a thermoplastic or a thermoset material and composites are referred to generically as glass reinforced plastic (GRP). The fibres are generally made of glass. The type of fibre used has an effect on the strength and ease of processing of the material. Short fibres give a moderately strong but tough matrix which can be injection moulded using short production times. Longer fibres increase the tensile strength but result in longer production times – a disadvantage when making mass-produced components. One of the most useful composites is glass reinforced polyester (GRP) which can be injection or compression moulded. Plastics can be alloyed together to give improved properties. A typical alloy will combine plastics which have good tensile and impact-resistance properties. Alloys are sometimes used for the matrix of GRP and also for plastics in their unreinforced state where strength and impact properties are not so critical.

Properties

Plastics are similar to steels in that it is not too difficult (using alloying, etc.) to modify the individual properties of a material. Changing one property however nearly always has an effect on other properties, hence it is the balance of properties that is important. From a design viewpoint, the properties of thermoplastics are the most important as they are in more common use. When considering the properties of thermoplastics, it is important to note the types of failure mode that affect them, for example:

- *Fatigue*: Whereas steel can fail by fatigue, a plastic (reinforced or unreinforced) reacts differently. Stress cycling generally causes crazing – this has the effect of whitening the appearance of the plastic and signifies that the structure has started to break down.
- *Weathering*: Polymers degrade when exposed to the environment; oxygen and ultraviolet light have the worst effects, particularly on thermoplastics. Weathering causes embrittlement and general degradation and can be considered analogous to the way that corrosion affects metals.

These apart, plastics exhibit the same types of properties as steel: tensile strength, stiffness, toughness, ductility, hardness, fatigue resistance, etc. The values of the mechanical properties of plastics, however, differ significantly from those of steel – this is an important point; it means that the design of plastic components must be conceived specifically with the use of plastics in mind. Material thicknesses and sections will be very different from those used for steel. Some general comparisons of properties are shown in Fig. 16.4.

Material	Tensile strength (MN/m ^{2})	Modulus (GN/m ^{2})	Density (kg/m ^{3})
Stainless steel	700–900	230	8
Low-carbon steel	300–500	200	4
Aluminium	170	70	2.7
ABS	50	3	1.04
Polycarbonate	60	3	1.25
Polycarbonate (30% glass-filled)	120	10	1.45
SMC	80	12	1.9
Polyester (filament wound)	800	50	2

Figure 16.4 Comparison of properties – plastic versus steel

Tensile strength   Unreinforced plastics have a much lower tensile strength than steel. The maximum ultimate tensile strength (UTS) that can be achieved by an unreinforced thermoplastic is about 120 MN/m^2 compared to 300 MN/m^2 for deep-drawing low-carbon steel. The modulus is also much lower: 3–4 GN/m^2 for thermoplastics and up to 100 GN/m^2 for thermosets compared to a steel value of about 200 GN/m^2. Glass fibre reinforcement can have a dramatic effect: an advanced filament-wound polyester can have a UTS of 1000–1200 MN/m^2 (comparable to high tensile steel) and a Young's modulus of 220 GN/m^2. The shear strength of plastics is important, particularly for car bodyshell applications where pure shear is one of the design stress cases. The shear performance of plastics follows similar principles to that of steel, with a direct relationship to the tensile strength.

Toughness   The toughness, or 'impact resistance', of plastics is poor – it is one of their inherent weak points. Impact resistance tends not to be measured quantitatively as in steels (there is no direct equivalent to the Charpy V-notch test, for instance). Instead, a rating letter A to D is used – a more qualitative method. One positive difference from steel is that the impact strength of plastic is not so sensitive to low temperatures. Some plastics, such as polycarbonate and its alloys, have been developed specifically to have high impact properties (safety hats and motorcycle crash helmets are made of polycarbonate) but the impact properties of other plastics are, at best, variable.

Creep Like steel, plastic will suffer from creep deformation under constant stress well below its yield point. The stress–strain performance is similar to steel, showing an elastic region at low strains followed by a non-linear deformation. The difference is that plastics are more susceptible to strain rate (and temperature) hence making a true value of Young's modulus difficult to specify – its behaviour is properly described as *visco-elastic*. Creep does occur though at stresses within the elastic region.

Chemical resistance This is another area in which plastics are very variable. The lower quality commodity thermoplastics are susceptible to attack by acids and alkalis but for engineering thermoplastics and thermosets, the problem has been largely overcome and they are resistant to many common chemicals.

You can see that there can be quite a lot of variability in important mechanical properties, across the range of plastic materials. As a very general rule, thermoplastics have a wider range of properties than do thermosets, but the balance of properties of individual thermosets is generally not so flexible as for steel. This means that they have more design constraints. Unreinforced plastics have limited application and need to be thicker than steel to do the same job. GRP composites such as SMC are the most useful, giving properties comparable to steel in some (but not all) cases.

16.5 Uses of plastics in cars

The use of plastics in most production cars is steadily increasing. From a total number of the 15 000 or so separate components that make up a car, about 700 are now made out of plastic, reducing its weight from approximately 1200 kg to 1000 kg. This means that nearly 100 kg of plastic is used in an average car (plastic is slightly less than half the weight of an equivalent steel section producing the same strength). Of this, about 75 kg is used for dashboard facia interior, trim and under-bonnet components, rather than load-bearing structural bodyshell parts. The figure varies between models and manufacturers but 75 kg (or 75 per cent of the total usage) is a good average figure. Structural bodyshell panels present the biggest technical challenge to the use of plastics from a design viewpoint. Perhaps for this reason, two separate methods of use have developed.

- True monocoque panels. Here, the plastic bodyshell panels alone replace the use of steel. They take the full structural and imposed loads, without any subframe reinforcement.
- Subframe supported. This is a compromise solution – a metal subframe is used, supporting either the full structure of the car, or sometimes only at the front end,

extending forward from the door shut-post. It is more common in two-door vehicle designs. The subframe is used to support the plastic panels, which are bonded (rather than bolted) to it, hence they are only partially load-bearing. The subframe also increases rigidity and impact resistance. This second method is a very different approach to the full structural use of plastic bodyshell panels – there is also less of an advantage in weight saving, unless the subframe is made out of a lighter material such as aluminium.

Before taking a more detailed look at specific plastic components it is worth thinking once again about the implications for production. Mass-produced car manufacture suffers from the problem that plastic parts, whether injection, blow or compression-moulded, take longer to produce than conventional stamped steel panels. A typical bodyshell consists of 25–30 separate panels and although this can be reduced by

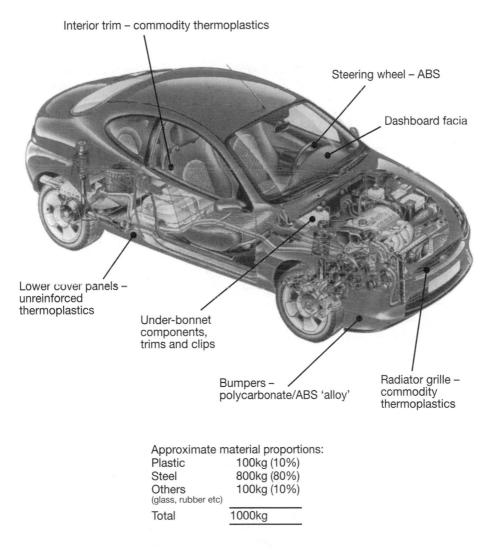

Interior trim – commodity thermoplastics

Steering wheel – ABS

Dashboard facia

Lower cover panels – unreinforced thermoplastics

Under-bonnet components, trims and clips

Bumpers – polycarbonate/ABS 'alloy'

Radiator grille – commodity thermoplastics

Approximate material proportions:

Plastic	100kg (10%)	
Steel	800kg (80%)	
Others (glass, rubber etc)	100kg (10%)	
Total	1000kg	

Figure 16.5 The use of plastics in mass-produced cars

about half for plastic components (because moulded plastic shapes can be more complex) the total production time is still longer. This is a pivotal commercial consideration, equally as important as the issue of mechanical properties.

Figure 16.5 shows the typical use of plastic materials in a modern mass-produced car. Some further details and the main reasons for the choice of material are given below.

Roof and underpan

These are key structural components, providing much of the strength and torsional resistance of the bodyshell. The only plastic material which is suitable is compression moulded SMC, a polyamide/polyester resin bonded with glass fibre. Other materials either are not strong enough or suffer badly from crazing due to the continual twisting imposed from the suspension and 'running gear'.

Sills

Sills are also an important structural member, transferring mainly torsional loads between the underpan and sides of the bodyshell. Technically, sills can be made from press-moulded SMC but are best moulded integral with the underpan – this gives better rigidity. There is always a problem with impact resistance – sills have to resist forces caused by side impacts and even the stronger varieties of SMC perform poorly without the support of a metal subframe. Side impact resistance is an important statutory requirement and must be confirmed by full-size crash tests before a car model can be licensed for sale and use.

Bumpers

Bumpers are also subject to legislative requirements – they must resist a low-speed frontal and rear impact test. This poses an interesting design problem; a plastic bumper must be strong and resistant to impact but also have flexibility to allow it to spring back to shape after a minor bump. The best material to provide these properties is an alloy of polycarbonate and ABS. Manufacturers such as Rover, Renault and Porsche have also used compression moulded SMC with some success. Fiat and Lancia have produced models with single-piece bumpers made from injection moulded polycarbonate with only 'trace' alloying constituents. Most production cars now use plastic bumpers but some do suffer from crazing, particularly around the mounting points. Cracking is also not unknown on older components (look in a car park).

Bonnets

Bonnets are, strictly, not load-bearing panels although they do assist the rigidity of the bodyshell when closed – it depends on the specific design. The key criterion is the ability of the bonnet to absorb frontal crash loads – it is part of the 'crumple-zone' that deforms, absorbs energy and helps protect the passenger compartment during a collision. Manufacturers of cars with highly-styled front ends (Citroën is a good example) use press-moulded SMC. Large bonnet panels need additional stiffening because single-thickness panels are too flexible.

Doors

Doors, like bonnets, are not strictly structural but do provide rigidity, when closed, to the bodyshell. They look suitable for the easy use of plastics but in practice they cause a few specific problems:

- *Hinge fixings*. Car door hinges take a heavy load and it is difficult to design a sufficiently strong hinge fixing in an all-plastic component. High stresses and impact loads (when the door is slammed shut) soon cause plastic fixings to weaken so additional steel shut plates and strengtheners, bonded to the door skin, are needed.
- *Side impact protection* (*SIP*). SIP is becoming increasingly important in car design owing to developing legislation. A single-skin door design has negligible SIP so a more rigid, cellular design is needed. This puts a constraint on the type of production method that can be used because several separate panels are needed to make up the door assembly.
- *Crazing*. Repeated opening and closing of the door causes crazing, particularly around the handles, catches and fixings. SMC or standard GRP composites are normally used – but these still have problems.

Several manufacturers (Fiat, Seat, etc.) fit cover panels, of unreinforced thermoplastic, over the steel door skin on the lower half of the door extending, on some models, to the front and rear quarter panels. The main purpose of these is resistance to stone-chips and corrosion – the doors still rely on their steel frame for structural strength.

Wings

Wings and valances are non-structural components. Their shape and application makes them particularly suitable for manufacture in GRP, the plastic matrix normally being polyurethane or polyamide. Rigidity can be a problem on those designs using flat-sided rather than bulbous shaped wings.

Tailgates

Plastic tailgates are used by several major manufacturers – SMC is suitable. The main design criterion is rigidity. If you look underneath the internal cover-trim of a plastic tailgate you will see that it has been moulded in two or three parts – a moulded box-section is frequently bonded around the tailgate rim to give some rigidity to the panels. Any stiffening effect of window glass is ignored for design purposes.

Fuel tanks

This is a controversial area. The simple shape of a petrol tank makes it suitable for rapid manufacture, in one piece, by the blow moulding process. Strength requirements are low and several types of plastic are available which have the necessary chemical resistance to petrol and diesel fuels. This is not a development that has yet been adopted by all major car manufacturers – the safety implications are still unclear.

Other components

Other, mainly non-structural, components that are made from plastic materials are:

- dashboards and facias (blended ABS);
- heater bodies and similar housings (various polyamide thermoplastics);
- radiator grilles (commodity thermoplastics);
- steering wheels (ABS);
- under-bonnet brackets and clips;
- interior trim items such as door cappings, consoles, handles, mirror frames, switches and electrical component housings.

From this brief outline it should be clear that the use of plastics for bodyshell components is not trouble-free. A further, generic problem of external bodyshell panels is the issue of surface stability and finish. External surfaces must have a perfectly smooth surface which will take and retain the paint finish. Early GRP suffered from serious paint flaking and cracking but most of these problems have been overcome in the past few years. Ideally, a plastic material can be self-coloured, so it can be used in its 'as moulded' condition, without further treatment.

16.6 Case study task

The case study task in this chapter is about developing the use of plastics in mass-production cars. We concluded, early in the chapter, that developments are driven (and sometimes constrained) by economy, recycleability and production cost considerations and that, on balance, these point to the increased use of plastic materials for bodyshell components. We are interested in the methodology that accompanies such developments in technology – the possible use of plastics for fully structural bodyshell components is an important innovative step that has to carry with it its own way of design thinking. To develop this point the case study task will take the important step of imposing a technology step. We will move from considering petrol driven vehicles to electric-powered ones. This has the effect of intensifying the design pressures – whereas with petrol vehicles, the use of plastics is a technical option, for electric vehicles it becomes a necessity due to weight limitations. There is also a technology aspect to it; mass-produced electric vehicles represent the main target of the world's motor manufacturers and it is generally considered that they will have to replace petrol-driven types over the next hundred years as easily convertible fossil fuels become depleted.

Your task is to formulate design features for a new and innovative electric car called the 'Citybee'. Outline technical information on the Citybee is provided, from which the following need to be decided:

- bodyshell materials;
- bodyshell (and individual panel) design features such as size, shape and method of fixing.

These areas need to take into account the real technical and production limitations of plastic materials. Decisions need to relate to a car design that is capable of mass-production, not one that can only be produced economically in small batches.

CITYBEE: OUTLINE DESIGN BRIEF

Basic data

Overall dimensions — length 2.75 m maximum
— width 1.5 m maximum
— height 1.7 m maximum

Ground clearance — 130 mm

Kerb weight — 900 kg (unloaded) including 250 kg battery pack

Capacity — two people with 30 kg luggage (max dimensions 1000 mm × 500 mm × 500 mm)

Power — Single electric motor capacity 19 kW

Performance — 160 km per charge. Maximum speed 60 kph

Construction

The bodyshell is of recycleable plastic and should be strong enough to resist normal operational loads. The Citybee will be subject to statutory crash tests – hence the bodyshell needs to have front and rear 'crumple zones' which will deform in a collision, absorbing the energy of impact. Some kind of protection also needs to be provided against side impacts. The size of metal subframe (if absolutely necessary) needs to be kept as small as possible so as to minimise weight and material costs.

Plastic bodyshell components must be suitable for manufacture using a moulding technique. The number of separate bodyshell components must be as few as possible, to minimise moulding production time. The Citybee does not need an opening bonnet; the battery pack and motor can be accessed via the rear tailgate.

The front and rear suspension both operate on the 'swinging arm' principle with independent mounting points to the subframe. The wheel/suspension/brake assemblies are conventional proprietary components as used on small petrol-driven cars.

Assembly

The bodyshell and subframe need to be separate (for repair and replacement purposes). A minimum of bolted connections should be used, for ease of assembly, but there should be no glued or bonded joints that prevent the bodyshell being easily removed after assembly. Bodyshell connections must, however, be secure so that they will not become loose during the operational life of the vehicle (they also need to be accessible for checking during the MOT test).

See also the Citybee concept design sketches and styling notes (Fig. 16.7).

Figure 16.6 Citybee outline design brief

What kind of problem is this? At first glance, it appears to be a complex nested problem containing all manner of technical and procedural problems mixed in together, with little indication as to the best place to start. Look more carefully at it though: most of the real design criteria of using plastics are related directly to the technical acceptability of plastic materials for the job that they need to do. The main one that isn't is the problem of *production method* – this may have production implications (it does) but it still has a firm technical basis. Do you see how production time constraints could be improved, or overcome, by optimising the

CITYBEE: STYLING NOTES

Here are some notes extracted from the 'concept design' styling report for the Citybee (see also the sketches below)

General shape
- Snubnosed contours (front engine space not required).
- 'Wrap around' panel at front to give smooth lines. Avoid sharp edges.
- Separate headlight pods.
- Maintain normal headroom height: so 'fastback' shape not feasible.
- Windscreen slope angle approx 45°. Vertical tailgate.
- 'Wrap around' rear panel incorporates tail light apertures.
- Roof panel almost flat with single curvature edges.
  Maximise door width (1100 mm minimum) for access.
- Recess all handles, etc. – no exterior bodyshell projections.
- Single colour bodyshell. Black or colour-matched front and rear bumpers.
- Smooth reflective finish equivalent to cellulose-painted steel.
- Flush window-glass and no protruding wheel arches (for smooth contour lines).

Points to avoid
- Too much of a 'slab sided' contour, especially at the front end.
- Looking like a Reliant Robin.

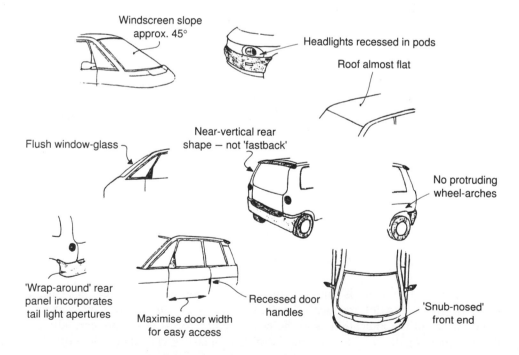

Figure 16.7 Citybee - styling notes

number of moulded panels that the body shell is made of – hence keeping the overall moulding time down? In essence, the design of plastic cars is a linear technical problem, so it will respond to the type of analysis outlined in Fig. 2.3 in Chapter 2. This can help your thinking.

Figures 16.6 and 16.7 give outline design information and styling notes about the Citybee electric car. This is currently at the stage of being an experimental vehicle produced in small batches of about 20 in a factory in Norway. They have been sold in small numbers to various local authorities worldwide on a trial basis for specific duties, mainly acting as public transport links to ferry passengers to and from regional rapid transit stations. Some have been sold for private use in congested city centres.

The exercise is to produce a design submission document for the Citybee bodyshell, proposing design features that maximise its suitability for mass production. To specify properly the design, the following items need to be included:

- outline sketches of the main bodyshell panel shapes and design features, to include the underpan/sills/side panel/roof assembly, doors, bonnet and tailgate as applicable to your design; show how the panels fit together;
- sketches showing the main tensile, shear and torsional stresses on the shell components;
- an inventory of the plastic components showing the type of plastic recommended;
- a beginning and an end to the document. It needs to look like an actual design submission that could be discussed with the Citybee manufacturers.

Within these constraints, the detailed format and content of the design submission is open to your own initiative. Remember though that, to be effective, it should address properly the relevant technical and production *problems*, as well as the advantages, that accompany the use of plastics.

16.7 Nomenclature

ABS   Acrylonitrile–butadiene–styrene. A commodity thermoplastic.
GRP   Glass reinforced plastic. A general term for a thermoplastic or thermoset matrix reinforced with glass fibres.
PC   Polycarbonate. An engineering thermoplastic.
SIP   Side impact protection.
SMC   Sheet moulding compound. A reinforced polyester.

17 Motorcycles – design and project costing

Keywords

Motorcycle design and manufacture – the whole vehicle concept (WVC) – research and development – engineering phase – prototype manufacture – production. Design project costing – basic financial concepts – project cost budgets – break-even analysis – P + L account – balance sheet – project and company-level financial records.

17.1 Objectives

This case study is about design and development costing, or to be more accurate, the relationships between the *activities* of designing and developing an engineering product and the way in which these activities are *costed*. It shows how to itemise and classify costs and how to fit them into established methods of management and financial accounting. These activities play a central role within all engineering and product disciplines that operate on a commercial basis. The case uses as an example the design and development of road-going motorcycles. The characteristics of the UK motorcycle industry provide a 'good fit' with the key messages of the case study – it is far from unique however – many other engineering sectors producing consumer or industrial goods follow a similar pattern.

17.2 The problem

There is a strong historical perspective to this case study. For nearly 50 years, UK motorcycle manufacturers were dominant in the world market for road-going machines. Companies such as Vincent, Rudge, Norton, BSA and Triumph all produced, at some time, brand-leading models. Technological innovations such as the disc brake, shaft drive and swinging-arm suspension were pioneered by these manufacturers for motorcycle use. Commercial success followed: in 1960 total UK output of motorcycles above 200 cc was nearly 50 000 units, representing 30 per cent of the world market.

It didn't last. From peak sales in 1962 to the closure of the final production line in 1978 (a sobering period of less than 5000 days) production declined to effectively zero. The sad truth was that the machine designs had become just too outdated – they were technically obsolete, and had been so for some time. The inquiries started, and ended, with various conclusions. Things were perhaps best summed up, however, by the words of one senior engineering director at the time:

... it was not that we were short of new design ideas, we had quite a few: OK, budgets were a problem, but we always had some money put aside, we just didn't seem to be able to put things together. Some design programmes cost too much, and had to be curtailed, whereas others appeared overfunded. Yes, it's a source of despair to me, years of so-called development but the models in the showroom stayed the same. And the design budgets – to this day I really don't know where some of this money actually *went*.

Therein lies the problem – design projects and their costs exist together, and they can be difficult to control.

17.3 Motorcycle design and manufacture

Motorcycle design, like the design of so many other things, invariably ends up as a compromise. Overt selling points such as a machine's performance and looks have to be combined with old, and some new, engineering practices. Both must be tempered by the practicalities and commercial realities of mass-production. The general pattern however is well defined – and surprisingly similar across international boundaries. The process is divided into the basic chronological steps shown in Fig. 17.1 and each step has costs associated with it.

Step 1: the whole vehicle concept (WVC)

As the very first element in the development of a new motorcycle design, this step is essentially a *creative* one. A manufacturer commissions two or three new designs (normally from an outside designer) with the objective of turning these into full-size vehicle 'mock-ups'. These are made of wood, fibreglass or plastic shells and are purely static 'models'. Although non-working they are finished, aesthetically, to a high level representative of the looks of the final motorcycle. They are known as *concept vehicles* (the new Triumph Daytona shown in Fig. 17.2 started life as a concept model). The purpose of concept vehicles is to stimulate the market (they are shown in the press and public exhibitions) and to obtain some customer feedback about the design. The manufacturer does not go to the extent of producing detailed engineering specifications, as such, but will display broad design parameters for the main systems. There are five main systems for a motorcycle: power unit, drive train, suspension/running gear, structure (frame) and 'fitted equipment'. Parameters such as performance, power/weight ratio and handling play a large part in the way that the concept vehicle is presented to the public, to gauge their interest.

WVC costs   Although only a few detailed engineering or production-line expenses are incurred in producing concept vehicles, there are still very real costs involved. They are:

- initial market research (should the concept vehicle be 125 cc or 1200 cc?);
- design commissions and artists' impressions;
- some brief technical research (engine types and powers, new frame designs, etc.);
- drawings of the mock-ups;
- construction for the mock-ups;

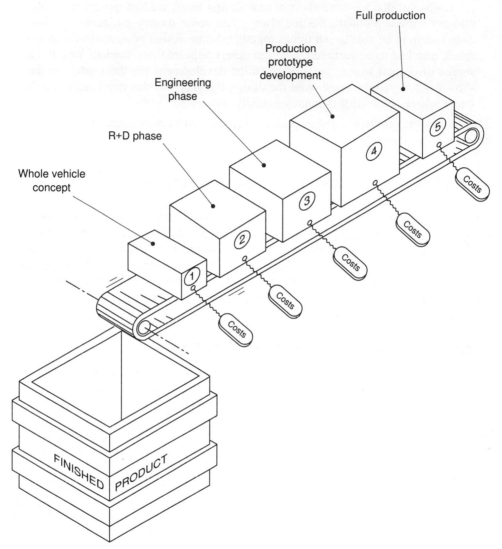

Figure 17.1 Five steps to a new design

- painting and graphics;
- advertising and exhibition costs (to gauge market response);
- the inevitable management and administration costs to organise all of this.

Note how most of the costs at this stage are 'doing costs' (*revenue expenditure*) rather than *capital* costs, i.e. related to buying items which can be classed as a permanent part of the manufacturer's business.

Step 2: the research and development phase

The research and development (R + D) phase is a series of preliminary activities necessary to turn the features of the concept vehicle into an engineering reality. At

Figure 17.2 The 1997 Triumph Daytona – started life as a 'concept model'

this stage, the idea of the 'whole vehicle concept' still applies, so R + D tends to be spread throughout the various engineering disciplines – note that it is *technical* R + D, related purely to technological aspects rather than the production/manufacturing process. It is often specialised and detailed. The main factor governing the effectiveness of R + D is the way that it is targeted. The available technical scope is wide, so it is normal for R + D activities to be apportioned along similar lines to the five main motorcycle 'systems'. These are, in very approximate order of their R + D priority:

- *Power unit.* Motorcycle engines are high-revving (10 000–15 000 rpm) with particular design features that differentiate them from production automobile models. The power-to-weight ratio is a key parameter, so significant R + D is performed on the use of lightweight alloys. Engine noise and vibration are important areas of quantitative and empirically based research. Owing to the complexity of engine technology, most motorcycle manufacturers rely heavily on highly specialised engine design consultancies.
- *Suspension and running gear.* Modern motorcycles are now extremely fast; speeds of 250–275 km/hr are easily achievable by the larger models so the braking system must be specially designed to cope. Features such as carbon fibre brakes, ventilated discs and anti-lock braking systems are under continuous development. Handling characteristics of a motorcycle change with such aspects as weight distribution, geometry and rigidity, so dynamic modelling is important for each new motorcycle design.
- *Structure.* The structure comprises the main frame, the rear swinging arm, and the front forks assemblies. The effects of frame geometry are very well known so little

fundamental research work is needed, except for specialised racing variations. Flexure and distortions *are* important however – this is a fairly standard finite element exercise using proprietary computer packages. Frames are available in several manufactured forms ranging from standard welded steel tubes to the use of high-strength alloys and cast construction.

- *Fitted equipment*. This includes the electrical items and the advanced electronics used for the engine management system. Many of the components in this category are proprietary 'subcontract' items so a motorcycle manufacturer tends not to have direct involvement on short-term model-specific R + D projects. Programmes are more often generic and longer-term.

- *Drive train*. Apart from basic size and strength considerations (engineering rather than R + D activities), fundamental R + D activity related to a model's drive train is unlikely. The four main technical options – chain drive, shaft drive, belt drive and fluid transmission – have all been used in production motorcycles and their characteristics are well understood.

A typical R + D programme will contain activities from some of these five systems. It is rare for all to be included – development does not usually proceed on all 'fronts' at once, for reasons of practicality and cost.

R + D costs R + D costs have two characteristics: they can be *substantial* and by their nature have a habit of being *intangible* and difficult to identify (more of this later). In the motorcycle industry, where vehicle production numbers are less than one tenth of that of automobiles, the 'unit cost' of R + D is consequently much higher. A typical R + D cost breakdown is as follows.

Management costs For low volume production, the issue of R + D management costs needs to be taken seriously. The specialised nature of the R + D and the fact that subcontracted consultancy companies are invariably used requires clear management direction if any results are to be achieved. Specific management time-costs which can be identified are:

- the project manager;
- in-house technical specialists – typically, input is required for each of the five main 'systems' described previously. Such costs tend to be incurred 'by department' within the manufacturer.

Subcontract specialists Specialist consultancy assistance relating to engine design and development normally constitutes the largest subcontract cost. Commissions of 3000–4000 hours are normal, a figure of around 1000 hours being generally accepted as the minimum input required to do something useful on a model-specific project. A typical breakdown will be:

- engine design consultancy (including noise and vibration);
- dynamic analysis consultancy;
- finite element stress analysis/aerodynamic consultancy.

The consultancy contracts will normally be organised on a consolidated 'time only' work hour basis, to include any computer costs incurred by the consultancy company.

In-house computer costs A manufacturer will invariably incur some in-house computer costs for integrating the input from subcontracted consultants and for studying some individual technical aspects of the motorcycle design. Costs can be divided into:

- hardware costs (purchase or lease arrangements);
- software purchase and maintenance costs;
- licence costs;
- operation (staff) costs.

Step 3: the engineering phase

The end result of the 'technological' R + D is to bring a sense of technical reality to bear on the 'concept vehicle' design. Invariably, the new motorcycle will have a lower innovative level than the concept vehicle. This is not so much an admission of technological defeat as an observation on the fact that concept vehicles are *meant* to be over-innovative – they can be thought of as an indivisible part of the creative process. The R + D outlined in step 2, although technical, is normally at a relatively high (although not quite 'generic') level and so most of the design of the new motorcycle still remains to be done. This comes during the engineering phase – the start of a series of practical steps that will result in the production of a new motorcycle model. It is easier to think of this engineering phase as comprising two disciplines: design engineering and production engineering, each with their own cost categories.

Design engineering Figure 17.3 shows a diagrammatic representation of the engineering design phase. The stages are approximately chronological and each one can be overlain on each of the five main motorcycle 'systems' described previously, giving a matrix arrangement:

- *Concept design*. This follows on from the concept vehicle design discussed previously, by bringing together practical engineering features for judgement.
- *Embodiment design*. Perhaps the key point about embodiment design is that it starts to consider, for the first time in the design process, general arrangement and layout drawings. The objective at this stage is to lay a firm foundation for the detailed design to follow.
- *Detailed design*. This is sometimes referred to as 'design for manufacture'. It uses, as its rationale, the fundamental principle that a detailed design drawing must contain all the relevant information for manufacture. Drawings are therefore in the format of *engineering drawings*, fully compliant with standards such as BS 308 and its international equivalents.

At this stage, these chronological design stages are almost totally model-specific; there is likely to be little cross-over with more general technical research issues.

Design engineering costs Even at this stage, when a new motorcycle design exists predominantly on paper (or screen), significant costs are involved. These are mainly 'revenue costs', broadly divided into:

- Design team costs. These are the staff costs involved as the manufacturer's *design resource*. Work-hour costs are incurred for the project manager, project engineer

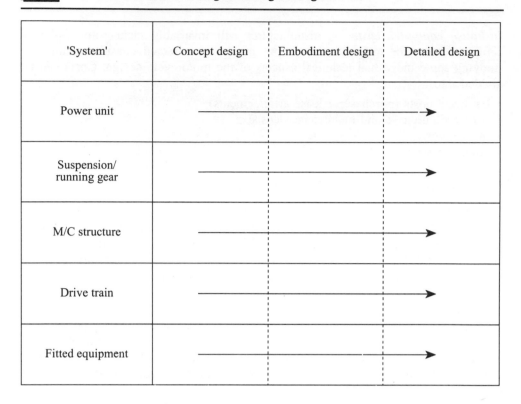

'System'	Concept design	Embodiment design	Detailed design
Power unit			
Suspension/ running gear			
M/C structure			
Drive train			
Fitted equipment			

Figure 17.3 The design phase 'matrix'

and technical specialists (three or four at this stage). Because of the complex nature of a new 'whole vehicle' design, the design team staff input is normally full-time.

- Drawing office costs. These are incurred whether traditional drafting or computer-aided design (CAD) is used – sometimes it is a combination of both methods. In contrast to design team input, drawing office work-hours are normally incurred 'part-time', i.e. several staff may be involved but they are unlikely to be dedicated to a single design project.
- Computer costs. Computer operating time costs are related to drawing office costs but tend to be recorded separately.
- Quality assurance (QA) costs. QA costs can be a controversial issue. Treat with care the notion that they do not actually represent a cost at all, i.e. that they can be seen as a negative cost (a credit). The reality is that the costs of a QA manager/ engineer and the implementation costs of a QA system are *real costs*, whether or not these costs may be recoverable by selling more, or better, motorcycles. If this is the case (it is difficult to prove, and frankly, impossible to predict) then any such savings will only manifest themselves during the sales phase. The engineering design phase will still be left considering the costs involved. Strictly, the engineering design phase is only involved with the earlier 'design stages' of a QA programme; most of the QA costs (perhaps 80–85 per cent) are incurred during the full production stage of manufacture.

One of the worst long-term mistakes that a company can make is to economise *too much* on design engineering costs – to try and pretend that they all, somehow, are a nicety, and not really necessary. Good engineering design is the key formative stage of a successful product; it is here that the fundamental issues of performance, appearance, reliability, safety, value and price are determined. Once a detailed design has been specified, then many of the future production costs have effectively already been set, so later changes may be difficult and mostly unsuccessful.

Production engineering   This follows on from design engineering as the second part of the engineering phase. Whereas design has a strong creative flavour, production engineering has the sober task of making available and assembling every one of the 10 000–15 000 individual components that make up the motorcycle. There can be very real *constraints* in production engineering, of both a technical and a logistic nature, and of cost. Production-line machines and tooling are very expensive – a single numerically controlled welding or machining unit can have a cost equivalent to the gross profit on 500 finished motorcycles – and there are many such machines in a production line. Luckily, many production line machines are flexible, particularly those that perform welding, pressing and machining operations, so they are adaptable to different tasks. Against this, however, lies the fact that machine technology does change very quickly. A machine seven or ten years old will be well behind the capability of a new machine in terms of machining speed, accuracy and flexibility.

The practice of assembling components is an important aspect of production engineering. Over the part 20 years, robotics has revolutionised vehicle manufacture, mainly for automobiles but also for the smaller production volumes applicable to motorcycles. The detailed design of the production line is a complex, multi-disciplinary exercise and several optimisation studies are often necessary to arrive at the best layout. The key issue, as always, is cost – most technical problems can be overcome, given sufficient time and funds, but cost is *always* subject to constraints. Detailed estimates of capital (tooling) costs, production line operation and staffing costs are undertaken at this stage.

Production engineering costs   Once again, costs are incurred before a firm design for the production line exists:

- Production planning costs. These comprise staff costs (full time) incurred by the manufacturer's production planning department. Planning involves quantitative exercises that may also incur computer costs (critical path/PERT analysis, etc.).
- Standardisation and QA costs.
- The inevitable management costs. Manufacturers tend to have highly structured production departments – the justification is normally something along the lines of 'production is what we do'. This means that lots of managers like to be involved, possibly more than are really necessary. Costs invariably follow.

Step 4: production prototype/development programme

This is really two steps seamed together. The first task of the new production line is to produce a batch (normally between 12 and 20) of the completed prototype machines. In practice, these are often not made entirely on the production line; up to 20 per cent of the vehicle may need to be hand-assembled in areas where the production line is not yet finished. The production prototypes are used for performance/reliability testing and safety tests (electrical safety, noise, emissions and crash testing) required by technical standards and statutory legislation. Special test rigs and tracks are used. At the same time, the production line is being finally completed. Some areas of change and development of the line are often necessary to address practical problems found. Although at this stage the production line has not started full-scale manufacture of motorcycles, most of the production line costs will have already been incurred. These include:

- production line capital costs (machine tools, jigs, etc.);
- revenue expenses for consumable tooling (moulds, dies, cutters);
- commissioning costs of the production line;
- trial runs;
- initial fault-finding and rectification;
- an initial proportion of the *variable costs of production.*

Step 5: the full production phase

There can be a surprisingly long delay between the manufacture of the pre-production prototypes and the point at which the assembly line moves into full production. There are commercial, marketing and sales aspects to this, as well as technical reasons. Once production starts, full volume production may not be instantaneous – it may take several weeks or months for minor 'snagging' problems to be rectified fully. One distinguishing feature of production line manufacture is that output is subject to constraint from the weakest, or slowest, link in the production 'chain'.

The costs of production follow a well-defined and established pattern, using the categories of fixed costs and variable costs (we will look at this apportionment in more detail later). Broadly, fixed costs are those which do not vary with the output volume (the number of motorcycles produced), such as factory overheads. Variable costs do vary with output volume. Common variable costs are:

- materials
- labour
- utilities (electricity, etc.)
- maintenance
- some of the QA-related costs.

These are broad categories only. We will see later that some of these costs may be defined in several ways, both for practicality and for accounting convenience.

17.4 Design projects: costing

Why bother accounting for costs? Yes, I'm afraid all the rumours are true, innovators really do not make good accountants – they are better at designing beans than counting them. Who wants to count beans anyway? As long as they're good technical beans, aesthetic beans, long-lasting and reliable beans, then bean buyers of the world will happily unite in buying them. And there is no reason why motorcycles should be *that* different. So where's the problem?

- The problem is NYM

Business is littered with the property of NYM – you could be forgiven for thinking that NYM is one of the precepts *of* commerce. NYM means: it is *not your money*. In the large and complete quantities necessary for engineering design and production, money is nearly always owned by someone else; they may be shareholders, investors, banks, finance houses or a combination of the four, but the effect is the same – *accountability* is required. This means that costs must be recorded, apportioned and generally treated in a way that other people can understand.

How do we do this? Fortunately, well-established methods have been developed. These methods have grown over several centuries. They are horribly imperfect but, frankly, are probably the best you are going to get. If a better financial system exists, then no-one has yet found it. The first thing to understand is that the system is built entirely on *convention*, so you may find it realistic, and matching your way of thinking; or you may not. The system is divided into many different levels and applications, which can be confusing, but it at least has the advantage of being absolutely consistent. You will not find, therefore, the technical contradictions and paradoxes that you will find in some other disciplines. Remember that this does not mean that *you* will find it realistic – it is still based on convention. The roots of this consistency lie with the key concepts that underlie all types of financially based systems. These concepts form an important part of the language of finance – we will look at them in turn.

Basic financial concepts

Costs From the viewpoint of engineering designers, the world of business finance looks different from that seen by, for instance, the sales and marketing fraternity. Engineering design is about *doing* – an activity which costs money. Sales and profit come later. Hence the best place to start is with those financial concepts that relate to costs. How is a *cost* defined? Here are two common attempts:

- a cost involves you paying money;
- a cost is payable in return for a benefit.

Wrong and wrong – there are multiple cases where neither of these statements is true. Here is a better one:

- A cost is something that you have to *account for*. You may or may not have to pay it, and it may or may not provide you with some benefit in return.

Easy definitions apart, there are many financial 'happenings' that go under the rather loose headings of 'cost'. There are five basic viewpoints which are relevant to design projects:

Cash costs This is where a clear cash debt is incurred as the result of a straightforward purchase of materials, labour, etc. It is payable within a short time – a few months. You will see cash costs referred to, rather loosely, as expenses. A cash cost is not necessarily paid in crisp notes; cheque or debit payments also qualify but it could be requested from you in cash. A cash cost, therefore, results in a net outflow of cash. It is this simple concept which causes many business problems. The final chapter of all bankrupt engineering companies reads much the same: 'Bills to pay but not enough cash available'.

Accounting costs These have almost the opposite characteristics of cash costs. Accounting costs are costs that, although they are represented as a cost within a company's financial system, are never intended to be settled in cash, i.e. they do not result in a cash outflow. They are however treated as a cost in *accounting terms* (i.e. as 'seen by' the accounts) and hence are subtracted from income, in the same ways as cash costs, when calculating *profit*. Typical accounting costs are company overheads, which have to be apportioned between different departments or projects of a company. Individual departmental and project managers, having had these costs allocated to them, do not necessarily see a direct benefit in proportion to the size of the cost.

Opportunity costs Opportunity costs are badly named; they could be more accurately described as 'opportunity losses of revenue' but rarely are. There is absolutely no cash transfer involved with an opportunity cost; it exists only in accounting terms. Basically, an opportunity cost is the revenue that you don't receive because your resources are busily employed doing something else, i.e. revenue missing because of a *lost opportunity*. For example, a skilled workman seconded from an important job where he can earn the company £Y per hour, to a lesser job where he can only generate £Z per hour, has an opportunity cost (to the lesser job) of £$(Y - Z)$ per hour. Because of the interminable practical arguments that can result from consideration of what all the labour force *could* be doing (but are not), you can see the awful intangibility that surrounds the concept of opportunity cost. For this reason, we will see later that its use is restricted to only a few parts of a company's financial system.

Fixed costs A fixed cost is one in which the size of the cost is independent of the volume of production (how many products are made). This does not mean that it is fixed, as such; just that if it does change, this is not due to the volume of production. Note the terminology here: units of production are traditionally referred to as *volume*. The only reason is convention. Just to complicate matters, fixed costs may have the properties of cash costs or accounting costs as previously described. For example:

- The cost of production line floor space apportioned to a project is a fixed cost (it is unrelated to production volume) but is also an accounting cost because it is not payable in cash terms.
- The cost of strategic spare parts for production line machines is a fixed cost (they are needed whatever the production volume). The spares must be bought from suppliers, so they also represent a cash cost.

Variable costs Variable costs are dependent on production volume – they may be directly proportional to the volume of production units or have a more complex, non-linear relationship. As with fixed costs, variable costs can have either cash cost or accounting cost characteristics. The tendency however is for them to be cash costs, owing to their tangibility. In production line manufacture, those labour and product material costs incurred directly by the production line are classed as variable costs.

Profit Profitability is a key indicator of financial performance. The term *profit margin* is well accepted as representing the amount that is left from sales revenue, once all the costs (of a cash and accountancy variety) have been taken out. Purely for accounting convenience, profit can be expressed in two ways:

- gross profit = sales revenue − cost of goods sold (COGS)
- net profit = sales revenue − (COGS + overheads + everything else).

Note that these are very different quantities – apart from that, about the only interesting fact is that gross profit is not adulterated by overheads and a myriad of (non-cash) accounting costs. It is therefore often a better determinant of the profitability of a production process, but remember that it still relies on the various conventions that have built up. You can see examples of gross and net profit later, in Fig. 17.8.

Assets Assets are things that are owned by a company in the course of carrying out its business activity. Most of them are material things like land, buildings, machines, materials and cash, but others, such as brand-names, may have less than an air of total tangibility about them. Simplistically, there are two types: fixed assets and current assets.

Fixed assets Fixed assets are resources, with a long economic life, that are used *to produce* goods or services within the business rather than for conversion themselves for resale. Land, buildings and machinery, as long as they are owned by the business, are the main examples. Fixed assets are subject to a mechanism of *depreciation* whereby their recorded 'book value' to the business is reduced, year by year, over the assets' working lives. This is also known as amortisation.

Current assets In contrast, current assets are not a permanent feature of the business. They are possessions like consumable tooling, materials and cash which can be expected to be either converted into cash, used up or spent, within a short time – normally taken as being 12 months. Something that is accounted for (conventions again, remember) as a current asset cannot be a fixed asset, and vice versa.

Net cash flow We have left the most important financial concept until almost last. The concept of net cash flow is one which is highly divisible – it can be applied to an individual project or department, as well as at the highest 'company level' of a business. It is a very simple concept:

- net cash flow = cash in − cash out.

Net cash flow can therefore be either a positive or negative quantity. It is generally taken that net cash flow is calculated over a period of time, which can vary, depending upon the timescale of the analysis. This means you can refer to the net cash flow of a *project*, or the net cash flow of a *department*, perhaps between its three or six month management accounting periods. A type of mild paranoia surrounds the results of net cash flow calculations in many businesses – negative cash flows are generally seen as being undesirable.

Sources of funds The final set of concepts are those used to describe the sources of business funds – we will see later how these are represented on a balance sheet. This is the money that actually *constitutes* the business. You will see the following discrete and tedious categories being referred to.

Share capital This represents money that has been 'paid in' by shareholders in return for a proportionate interest in the company. There can be several types: *ordinary* shares (where the holder has 'normal' shareholding rights) or *preference* shares, in which the holder has preferential rights to a share of the profits (termed 'dividend'). Share capital, although it represents a register of various interests in the company, does not represent a 'debt' as such because it does not have to be repaid.

Long-term liabilities Long-term liabilities (LTL) represent money that has effectively been lent to the company. It therefore has to be paid back but the debt does not fall due for some time, as the title suggests. Once again there are several types:

- Debentures: a debenture is a long-term secured loan providing for a fixed rate of interest on the sum loaned, and specifying when the original amount (the principal) has to be paid back. A debenture holder is therefore not a member of the company – like a shareholder – but is a *creditor*, with a future financial claim against the business.
- Long-term loans: these are similar to debentures except that the loan may be unsecured, under which conditions the borrowed money is known as *loan stock*. The lender is a creditor of the company and is entitled to interest and repayment of the principal over the longer term.

Current liabilities You may find it confusing that 'sources' of funds can be classified as liabilities, but this is the convention. Current liabilities represent money that the company has, but that must be repaid within 12 months. You can think of it as being 'short-term capital'. There are several sources of short-term capital available to companies:

- *Overdraft*: commercial banks provide short-term funds (termed *working capital*) to finance the running of a company. It may take the form of loans repayable at short notice or by overdraft facilities. Overdrafts are repayable on demand – a depressing process known as *foreclosure*.
- *Trade credit*: trade credit is money advanced to a company by that company's suppliers, by their not requiring payment until some time after they have supplied goods or services. For small companies this is a useful form of interest-free short-term finance.
- *Debt factoring*: debt factoring is a way of passing over the financing of money owed *to* the company (debtors) to a special factoring or finance company. Hence the selling company is paid a percentage of their invoices in advance, before their

customers have actually paid anything. Debt factoring carries high interest rates, but it is a viable source of short-term funds.

These then are the basic financial concepts that lie behind common systems of costing and financial analysis. Taken individually, they are little more than shots in the dark – to make them meaningful they must be knitted together into a more complete system; we can then start to use them to represent real project and company situations. For this we need financial *pictures*.

Financial pictures

Pictures of what? A simple answer: pictures of the project, department or company that has to be expressed financially (remember the problem of NYM?). It is absolutely impossible for even the simplest financial situation to be expressed fully on a single chart or table. This is because of the nature of finance itself, and the existence of that famous phrase (so beloved by accountants) 'the temporal context'. All this means is that financial happenings occur over time – the situation is dynamic – which causes problems when you try to represent it on paper. The situation starts to become clearer: if you want to get a true picture of a financial situation, it is necessary to get a composite view – to look at it in several different ways. Hence various financial *representations* have been developed which will do this – we will look at some of the more important ones, in the context of applying them to engineering design situations. They exist at two levels: the project level and company level.

Pictures at the project level   Financial pictures at the project level are called 'management accounts', 'budgets' or something similar. Their common character-istic is that they have *consistency* only at the level of an individual project, i.e. the figures will only match up within the project-level calculations. This does not mean that project-level calculations do not have *relevance* outside the confines of the project (they will do, as they contribute to the overall financial situation of the business), but simply that the cumulative consequences of all the different project-level accounts and budgets are not necessarily considered when they are being prepared. Adjudication is often reserved until later, at a high level in the organisation. In general, project-level accounts do not attempt to specify *where* any funds are going to come from, other than to infer that they will be required. The three main representations (remember that representations are *all* that they are) are the project cost budget, the project cash requirement and the break-even analysis.

The project cost budget   This is perhaps the crudest of all the financial pictures. It is a simple list or schedule of all the costs involved in a project over its duration. It can be compiled retrospectively, but is most often used to try to provide a prediction of how much a project will cost. The level of detail in the list can vary enormously – it can include only a broad list of cost areas, such as 'design', 'production', etc., or it may be broken down in a more detailed way, identifying the costs of individual departments and parts of the project. Figure 17.4 shows a broad pre-production breakdown typical of a motorcycle design project. Project cost budgets tend to have the following distinguishing features:

Stage	£K
(1) Whole vehicle concept	30
(2) Research and development	125
(3) Engineering	800
(4) Production prototype/development	2200
(5) Production	(see B/E analysis)
Total fixed cost (pre-production)	£3155K

Figure 17.4 Project cost budget – example

- They assume a given 'output' for the project, such as type of design.
- The list of costs is normally expressed chronologically, in the rough order in which the costs will be incurred. This is mainly for convenience.
- No attempt is made to *account for* the costs, i.e. to explain where they come from (or when they will be paid back).
- Factors such as inflation, escalation and other commercial practicalities are conveniently ignored.
- 'Accounting' costs, which do not involve an actual cash transfer, *are* included.

The project cash requirement This is sometimes called a cash-flow budget, which is a little misleading, as it is still concerned only with costs that will be incurred, and does not yet consider any revenue that will be received once a project is complete. The project cash requirement is developed from the project cost budget but has several differences:

- It shows the cash requirement *over time*, normally on a monthly basis. Each set of monthly figures follows on from the previous one.
- Only figures that involve an actual cash transfer are included; those that represent a pure 'accounting cost' arc not.
- A fair level of cost breakdown is required in order that the costs themselves can be stated with some degree of certainty. This is in contrast to the project cost budget where groups of costs are often amalgamated together and expressed as a single approximate figure.

Figure 17.5 shows the project cash requirement (again on a pre-production basis) corresponding to the specimen project cost budget shown in Fig. 17.4. The difference in the total is caused by the neutral effect of 'accounting costs' as previously described. Note the significant effect that these have on the final apparent total – so it can be misleading to look only at a project cost budget *or* a project cash requirement; you need both to provide a separate but complementary viewpoint on the situation.

The break-even analysis The two previous pictures provide information purely on costs. To this we must add some information about the potential upside of our project. We have to consider a picture of *profitability* – the term that constitutes sweet music to generations of managers and directors. Together, the three pictures will provide a good composite view of the financial territory of a project. The easiest way to express profitability is to use break-even (B/E) analysis. You may also see this referred to as a cost volume profit (CVP) analysis. The analysis shows how cost and revenue (and therefore total profit) change with the production volume – so it is only relevant to batch or mass-produced products. It can be displayed as a matrix of figures but is easier to understand in the form of a graph, as shown in Fig. 17.6.

The horizontal axis indicates the volume of production, normally in hundreds or thousands of units – it also, indirectly, represents *time* as the production process extends over an extended timescale. The total cost of producing the product units is the combination of the fixed and variable costs, concepts described previously. Revenue from sales of the product, assumed to be proportional to the number of units made (and sold), is represented as line (SR) on the same chart. The position (X) at which the total cost and sales revenue lines intersect is known as the break-even point (BEP). After that, the project starts to generate profit. The BEP can be calculated arithmetically by:

$$BEP = \frac{Total\ fixed\ costs}{Sales\ income\ per\ unit - variable\ cost\ per\ unit}$$

Beyond the BEP, the fixed costs remain the same so total profit is increased each time an additional unit is produced and sold. This increase in profit per unit sold is known as the *marginal contribution*. Figure 17.6 shows a B/E situation consistent with the cost breakdown used for the project cost budget example in Fig. 17.4. The BEP is 868 units. It is important not to get too excited about the accuracy of a B/E analysis; broad assumptions have to be made in assessing whether costs are variable or fixed and also about the consistency of selling price as sales volume increases. This means that the results of B/E analyses should be treated as indicative, rather than truly prescriptive.

Pictures at the company level

Financial pictures of the project level are fine; their only problem is that they do not look at the knock-on effects that they can have on the overall situation and performance of the company of which they form a part. For project proposals to find acceptance at the higher 'empowered' levels of a company, they need to be coded to fit in with the financial scenery at that level. We can refer to this broadly as the *company level*. This is a brave step to take – whereas 'management accounts' used at the project level are rather informal, those at the company level (known as 'financial accounts') are very formalised, and follow well-defined patterns. One reason for this is that they are subject to external scrutiny for tax assessment purposes, and hence must be put together using known practices and conventions. The overall philosophy is that the financial pictures must represent *a true and fair view* of the financial situation as it is. It is important, therefore, to link up project proposals to these company-level financial pictures – this is part of the secret of presenting project proposals that have a chance of being accepted by companies' senior management. The situation is shown diagrammatically in Fig. 17.7. There are two items that are important:

The profit and loss (P + L) account The P + L account is easier to understand than its near neighbour, the balance sheet. The P + L account compares revenues and

	J	F	M	A	M	J	J	A	S	O	N	D	Total £'000
Whole vehicle concept													
Market research (c)	5	5											10
Design commissions (c)		5	5										10
Mock-ups (nc)			5	5									10
													30
Research and development													
R + D Management (nc)		5	10	10	10	10							45
Subcontracts (c)			10	10	10	10	10						50
Computing hardware (nc)			5	5									10
Computing software (c)				10	10								20
													125
Engineering													
Design team (nc)						100	100	100	100				400
Drawing office (nc)						20	20	20	20	20			100
Computing (nc)						20	20	20	20	20			100
QA (nc)					20	10	10	10	10	10	10	10	100
Production planning (nc)								20	20	20	20	20	100
													800

Prototype development

	1	2	3	4	5	6	7	8	9	10	11	12	Total
Production line (capital) (c)									600	600	600		1800
Production line (revenue) (c)									50	50	50		200
Production line (commissioning) (nc)								50	50	100	50		**200**
													2200
TOTAL	(5)	(10)	(15)	(20)	(20)	(10)	(10)	(20)	(670)	(670)	(670)	(70)	
CARRIED FWD	0	(5)	(15)	(30)	(50)	(70)	(80)	(90)	(110)	(780)	(1450)	(2120)	
TOTAL CASH REQUIREMENTS	(5)	(15)	(30)	(50)	(70)	(80)	(90)	(110)	(780)	(1450)	(2120)	(2190)	

(c) = cash costs   (nc) = non-cash costs   all costs in £000

Figure 17.5 Project cash flow requirement (pre-production) – example
(non-cash costs (shown shaded) do not form part of the cumulative cash requirement figures)

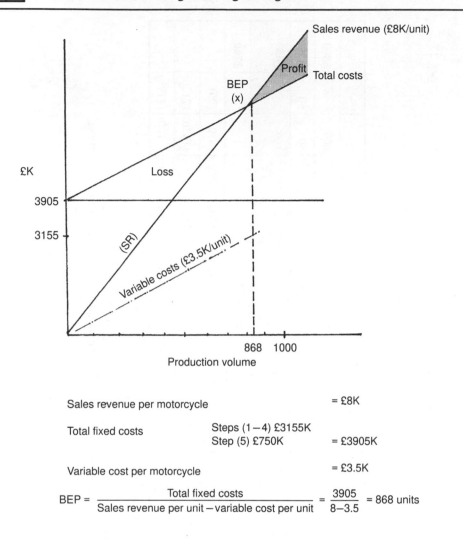

Figure 17.6 Break-even analysis – example

expenses (costs) over a fixed period, normally 12 months. It is also one of the financial records that is included in a company's published accounts. Figure 17.8 shows a sample structure and format for a company P + L account. The various headings should be taken as typical rather than definitive – many of the cost and overhead categories can be further subdivided. Note two main points:

- The P + L is *not cash-based*; some of the costs are 'accounting costs'. The major one in this category is depreciation.
- Note the way that profit is expressed in several forms: gross profit, operating profit before interest and tax (PBIT), and net profit. This is part of the system of conventions used for published company accounts. The final figure at the bottom of the account represents the net profit, after all deductions, that has been earned by the company over the 12-month period covered by the account.

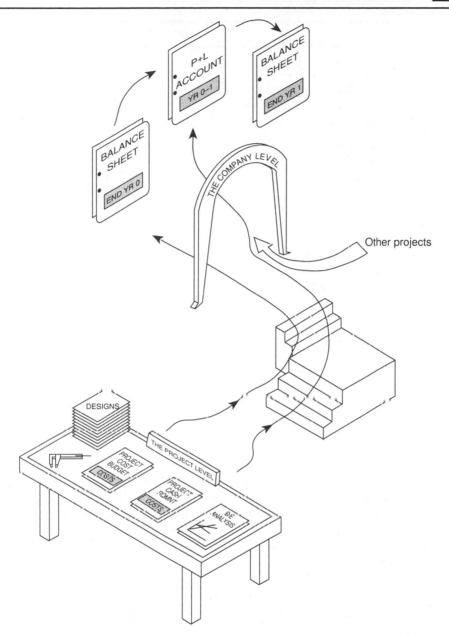

Figure 17.7 Links between 'project level' and 'company level'

Now for the big question: where does all net profit *go*? Its destination is the balance sheet, a separate financial 'picture' but one which is linked firmly to the P + L account. Net profit is transferred to the balance sheet (describing the situation at the end of the P + L period) into a category called *retained earnings* or sometimes *reserves*. Note that this does not mean that retained earnings exist in the form of cash – it is an 'accounting' quantity. It provides a first approximation but it is *not* a measure of 'how rich the company is'– many bankrupt companies show healthy

	£K	£K
Sales revenue		3000
Direct materials	400	
Direct labour	500	
Cost of goods sold (COGS)	900	(900)
Gross profit		2100
Overheads		
Depreciation	40	
Management	160	
Salaries	250	
Other fixed costs	300	
Utilities, etc.	70	
	820	(820)
Operating profit (PBIT)		1280
Interest	80	(80)
Profit before tax		1200
Corporation tax	300	(300)
Dividend	10	(10)
Net profit		890

Figure 17.8 P+L account – an example

reserves on their balance sheets. The P + L then supplies useful but *limited* information about the financial situation at company level; more information is available from the balance sheet.

The balance sheet The balance sheet is more difficult to understand. First and foremost, it is built around the concept of financial *balance*. Take a first look at Fig. 17.9 and note how the totals at the bottom of the left- and right-hand sides are identical. Balance sheets are all like this – the reason why this is so introduces one of the key principles of financial accounting, that of *duality*. The idea behind the dual aspect is that every financial transaction that takes place within a company produces two effects, and therefore has to be recorded twice. Looking again at Fig. 17.9 you can see this in action – the left-hand side (LHS) of the balance sheet shows crudely where money *is*, whilst the right-hand side (RHS) shows where it has come *from*. For example:

- When a company borrows money it appears on the LHS. At the same time, the borrowing creates a liability to the lender, which appears on the RHS.
- If an asset is bought, this appears on the LHS. The liability to pay for the asset comes with it, and shows as an equivalent sum on the RHS.

This means that the *assets of* a company are always balanced by the *claims on* that company. Bearing in mind this principle, the balance sheet is prepared to represent the situation at a single snapshot in time – it does not represent the situation over a period of time, as does the P + L account. Normally, a balance sheet will apply to the situation on that day at the end of a 12 month P + L period, so by definition, a

Fixed assets	£K		Long-term liabilities	£K
Land and buildings	275		Debentures	150
Plant and machinery	315			
	590	590		
Current assets			**Current liabilities**	
Stocks – raw materials	82		Trade creditors	150
Work in progress	124		Taxation	300
Finished goods	123		Dividend	10
		329		
			Sources of funds	
Trade debtors	357		Ordinary shares	220
Investments	50		Preference shares	298
Cash in bank	17		Retained earnings	215
		424		
Total assets		**1343**	**Total claims**	**1343**

Figure 17.9 Balance sheet – an example

balance sheet will change day by day. Figure 17.9 shows the categories of assets and sources of funds that appear on a balance sheet. These categories do not differ much; they are largely fixed by convention.

Together, the system of three project-level financial pictures – project cost breakdown, cash requirement and B/E analysis – can be combined within the two main company-level pictures – P + L account and balance-sheet to provide a true dynamic model of a company. This system has general application to all types of businesses but is particularly relevant to those that design, develop and manufacture engineering products.

17.5 Case study tasks

Rewind: autumn 2005 – the problem of project 'Z'

'So, just how much is this new motorcycle design going to cost?' The Managing Director continued, 'We've got CAPEX (capital cost) estimates here of £2 million for the new production line, and then there'll be the additional labour, all to build something that's not even designed yet. Seems like speculation to me; anyway, I don't see what's the matter with our 900 cc single cylinder Silver Spook model – built to *standards*, not for speed.'

'Design moves on sir, it's fluid, dynamic, iterative, all of these things. Change or die – the icy blast of competition sir, the . . .'

'That's the problem with you sales people; how much could we sell these new bikes for anyway: five, six perhaps?'

'Seven thousand pounds by my guess, and we could make 800 units per year – sir –'

'And the costs? – presumably the £2 million is just the tip of an infinite financial iceberg. Anyone got any cost estimates?'

(Design engineer, rustling Figs 17.10 and 17.11) 'Here they are, all the estimates from our in-house departments and subcontractors.'

'Can't present those to the shareholders, they're too random, and don't show the effect on the rest of the company finances. What we need are proper budgets.'

(Design engineer) 'What we need are proper budgets.'

(Sales engineer) 'I think what we need are proper budgets.'

(All) 'What we need are proper budgets.'

Task 1

Figures 17.10 and 17.11 show raw cost estimate information that has been gathered as part of your company's studies into a new and quite innovative motorcycle, known only as 'Project Z'. This motorcycle is really quite different from those that the company already produces – all five main systems, power unit, suspension, structure, drive train and fitted equipment, have innovative features. It has exceptional performance, being able to accelerate to a top speed of 280 km/hr in under ten seconds (it is a road-going machine, however, for which the statutory speed limit remains stubbornly at 112 km/hr).

As design engineer, you are convinced of the technical merits of the new machine but it is also your task to present the financial case for this design project in a coherent (and convincing) way. You must do this using accepted financial techniques and in a way that fits in with the characteristics and practices of the way that real industrial products are designed and manufactured – both of these areas are covered by the case study text. You will need to read the full text through to get a general feel for the situation, referring to individual parts in more detail as necessary. Don't forget the figures; they show important principles and cost information.

Then carry out the following tasks:

- Prepare a project cost budget for the new design using the information in Fig. 17.10. Costs should be in a logical order, recognising the five steps discussed in the case study text.
- Prepare a B/E chart for the production phase.
- Prepare a draft project cash requirement for the 12-month period from conception to the start of production. To do this you will have to make assumptions about the timing of the expenditure over the period – these assumptions are up to you but they must be clear and consistent with the broad chronology of the various cost steps as described in the text and in the sample Fig. 17.5.

Task 2

The next task is to predict the *effects* that the 'Project Z' cost budget will have on the overall financial position of the company.

- Assuming the sample P + L account given in Fig. 17.8 covers year 0 (the year before the design project *starts*), produce a predicted P + L account for:

DESIGN COSTS	£K
M/cycle concept design	7
Design commissions/artists	5
Engine design consultants	10
Construction of mock-ups	9
M/c detail design	20
Dynamic analysis:	
running gear	12
drive train	8
Embodiment design	6
F.E. consultancy	4
Design of mock-ups	4
TOTAL	85

MISCELLANEOUS COSTS	£K
Product line trials	24
WVC–tech. research	7
Graphic design	3
Engineering computer	9
In-house computer (R+D)	13
Standardisation	4
Prod. line revenue costs	210
Design exhibitions	3
Concept m/c – publicity	4
R+D contract – drive train	8
– fitted equip.	6
TOTAL	291

STAFF COSTS	£K
Production Manager	3
Drg Office Manager	12
Draughtsmen	3
Design team (technical)	21
Prod. planning eng.	4
Prod. line electricians	10
QA manager	4
Design manager	7
Design administrator	2
Standards engineer	7
QA engineer	7
R+D project managers	40
Prod. line commissioning	10
TOTAL	130

PRODUCTION ESTIMATES	£K
Variable material costs per unit	1.8
Variable labour costs per unit	0.8
Production line maintenance costs (fixed)	80
QA costs (fixed)	63
QA variable costs per unit	0.4

Figure 17.10 Project 'Z' initial cost estimates

– year 1 (when the design process is happening)
– year 2 (the first year of production).

Supplementary financial information about the project is given in Fig. 17.11. Assume that all other financial aspects of the business, apart from the existence of the new motorcycle design, remain the same.

1. Funding

It has been agreed that capital expenditure on fixed assets (£2000K) will be funded by taking out a long-term loan at 5% interest per annum. There will be an interest 'holiday' until the start of the second full year of production of the new motorcycle. Development costs of £526K will be funded by £500K of debenture stock at a simple interest rate of 6%, taken out at the beginning of year 1. The remainder will be funded by overdraft.

2. Depreciation

Depreciation on existing plant is linear, at £40K p.a. The new plant fixed assets will be depreciated at 10% p.a. starting from year 2.

3. Dividend

No shareholders' dividend will be paid in year 1 – this will revert to a payment of 0.83% of PBT during year 2.

Figure 17.11 Project 'Z' additional financial information

Task 3 (optional)

Assuming that the sample balance sheet given in Fig. 17.9 represents the position of the company at the end of year 0, produce a predicted balance sheet for the end of year 1 to show the effect of the new motorcycle project expenditure on the business. Use the supplementary information in Fig. 17.11 and, again, assume that all other financial aspects of the business remain constant.

18 Flue gas desulphurisation – total design

Keywords

What is 'total design'? – engineering, technology and context problems. Flue gas desulphurisation (FGD) technologies – the three main FGD processes – performance, reagents and by-products – technical problems – lifecycle costs. Making a choice using the 'total design model' – assessment reports.

18.1 Objectives

The objective of this case study is to show engineering design issues on their largest scale. Engineering projects to build chemical plants, power stations, mines and transport infrastructure are expensive and often controversial. Complex decisions must be made, involving multiple factors and uncertainties. Engineering design decisions form an important part of the process so the content of the case study has been chosen to illustrate this. It will also show you the complexity of some of the design issues present in large capital projects and the way in which they fit together (and sometimes oppose each other). The case study task is designed as a group exercise to help you become familiar with some important principles whilst gaining an appreciation of the technical aspects of the case study. It is also intended to help you understand the concept of *total design*.

'Total design' is a term used to describe that approach which gives the widest possible view of a design problem. It puts the engineering aspects of a design in context with the broader issues that surround it – issues that could not be seen by looking at the problem in a restricted way. Total design is also about organising yourself to approach the design task in a disciplined and structured way – and applying engineering and technical rigour, where necessary during the design process. The concept of total design as defined by Pugh (ref. 1. of Chapter 1) involves consideration, during a product's design process, of the characteristics of the market for that product. Large process plants, like flue gas desulphurisation, that have an environmental role (rather than producing a desirable chemical product) have a very indistinct market, perhaps better described by referring to it as a context. This can be expressed simply:

Total design = The technology + The engineering + The context.

By dividing the definition in this way we can see that the total design process (it is a *process*) is made up of technical and procedural parts. Neither the technological nor the engineering aspects of a design can stand alone on this large scale; they lie within a *context* which is governed by established procedure – the way that engineering projects *happen*. Total design, as well as being multidisciplinary, is multivariate, i.e. it has to include different viewpoints, which can make it complicated.

This study uses the case of flue gas desulphurisation (FGD), an emission control technology for large power stations. Fossil fired (coal and oil) power stations produce sulphur dioxide (SO_2) in their flue gases which combine with other chemicals in the atmosphere to form acid rain. There is now a worldwide trend, driven by environmental legislation, to reduce SO_2 emissions. This requires large-scale investment in new and developing technologies which have to overcome difficult process and engineering problems, against a complex background of statutory requirements and commercial pressures. It is a good illustration of the need for (and advantages of) a total design approach.

18.2 The problem

In the 1980s the UK emitted over 3.75 million tonnes of sulphur into the atmosphere, of which approximately 70 per cent came from fossil fired power stations. The EC Large Combustion Plant Directive requires that SO_2 emissions be reduced by 60 per cent from 1988 levels and this must be achieved by the year 2003. The situation is similar in other European countries. This means that FGD equipment has to be fitted to many power stations; a typical installation is shown in Fig. 18.1. Some other countries in the world have previously been in the same situation themselves and results were not good. Enthusiastic promises and hurried technical choices were made, accompanied by ever-increasing bills, as electricity utilities embarked on large-scale FGD retrofit of their existing power stations. Then came the problems; many plants failed to meet their performance guarantees and could not remove enough SO_2. Others either corroded badly or suffered process problems within the first 18 months of operation and some, sadly, did not work at all.

With hindsight, the problem with FGD projects divides into the following three parts.

Diverse technologies

There are many different FGD technologies – four major classes, each with eight or nine variants. Those with previous commercial application are owned by their licensors – large chemical engineering companies – whilst others exist only at the laboratory or pilot plant stage. The main FGD technology classes use different chemical reagents to remove the SO_2 from the flue gas and produce different by-products. All make enthusiastic claims about their desulphurisation efficiency but the large number of technology variants, and different applications, makes them difficult to verify.

Technical problems

Many FGD processes suffer from process problems. In some, the various sets of reactions have a tendency to 'block' each other in an unpredictable way whilst others suffer from scaling – again unpredictable – due to the presence of supersaturated

Figure 18.1 The FGD plant on Drax power station
(courtesy National Power plc)

solutions of sulphur- and calcium-based compounds. Materials of construction are also a problem; the process conditions inside the desulphurisation vessels are aggressive, causing rapid corrosion and erosion. High-nickel alloys and special plastics have to be used for some items of equipment.

Costs

FGD plants are all about *cost*; they cannot be operated at a profit, in the true commercial sense. The capital cost varies with the type of technology and can be up to £300m at 1997 prices to equip a large power station. Running costs have to be paid for the reagents, staffing and maintenance and there is a reduction in the power station's electrical output, which reduces its overall 'sent-out' efficiency and hence increases the generation cost of each unit of electricity. The income possible from selling the by-products of the FGD process is small, less than 20 per cent of the total running cost. Capital and running costs combine to give a *life-cycle cost* for the FGD plant and this varies significantly between technologies.

The combination of these three characteristics makes the choice of FGD plant for a particular application difficult. It is easy, as has already been proved, to get it wrong. So the problem is about how to make a rational choice, considering all the different arguments, whilst under the pressure of tight timescales to 'get the job done'. This is where the concept of total design can help. We will start with Fig. 18.2 which summarises the inherent problems of FGD projects in a form which fits in with the total design approach. First, however, we need to look at the technical aspects.

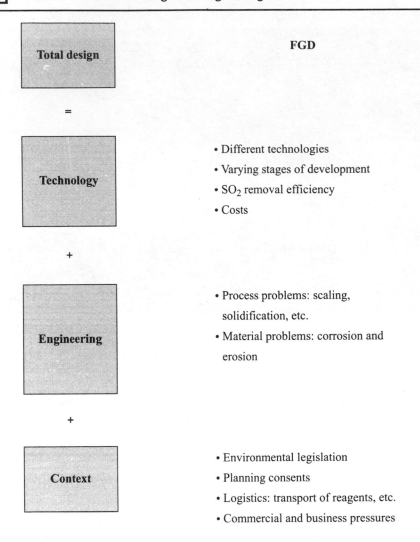

Figure 18.2 FGD problems – a summary

18.3 FGD: technical outline

Of all the available FGD technologies, three are of most importance, having developed to a state of commercial operation in several countries. These are the wet limestone–gypsum process, the lime spray-dry process and the Wellman–Lord regenerative process. We will look at them in turn – Fig. 18.3 shows the process flow diagrams.

The limestone–gypsum process

This process has been in commercial use since the 1970s. It uses limestone ($CaCO_3$) as a reagent and produces $CaSO_4$ (gypsum) as a main by-product. The limestone–

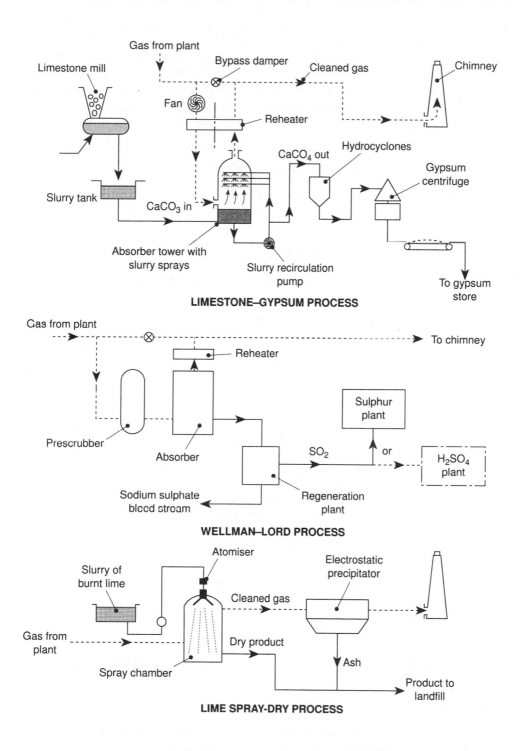

Figure 18.3 FGD technologies – process flow diagrams

gypsum process has several variants and all have the advantage of a high desulphurisation efficiency, up to 95 per cent removal, depending on the type of coal and limestone being used. The disadvantage is the high electrical power consumption required to operate the FGD plant – about 3–4 per cent of the power station's rated output. The flue gas is mixed with ground limestone slurry in an absorber, the resulting reacting liquor then being recirculated by a pumping circuit to increase the residence time. The $CaCO_3$ reacts with the SO_2 in the flue gas to produce calcium sulphite ($CaSO_3$):

$$CaCO_3 + SO_2 = CaSO_3 + CO_2$$

Air is blown, or *sparged*, into a tank at the bottom of the absorber tower. This completes the reaction by oxidising the $CaSO_3$ to $CaSO_4$ (gypsum). Both sets of reactions are sensitive to pH level:

$$2CaSO_3 + O_2 + 4H_2O = 2CaSO_4 + 4H_2O.$$

The gypsum solution is continuously bled-off from the pumping circuit and dewatered using cyclones and centrifuges to produce dry gypsum. Other, smaller bleed streams extract and treat other impurities (mainly CaCl) in the process liquor. The gypsum product can be used to make commercial plaster or wallboard.

The process is not immune from technical problems. Hydrochloric acid (HCl) originating from the coal is present in the flue gas and can interfere with the main desulphurisation reactions – a separate prescrubber vessel may be needed upstream of the absorber tower to remove the HCl. There is also the need to reheat the clean gas after it leaves the absorber (in order to increase the buoyancy of the chimney plume) as the wet spraying process reduces the temperature of the gas to below about 100°C. This requires regenerative reheaters which heat the cleaned gas by recovering the heat from the dirty gas before it enters the absorber. The process conditions in a limestone–gypsum FGD plant are very aggressive. The main slurry loops are alkaline (pH 1) and have a high concentration of suspended solids – up to 40 per cent. Normal steels cannot withstand such conditions so special materials of construction are needed such as high nickel alloys, glass-flake reinforced vinyl esters, and special plastics. The absorber towers are susceptible to plugging and scaling caused by the very low solubility of calcium and sulphur salts; hard scale of one metre thickness has been recorded in some plants.

The running costs of limestone–gypsum FGD plants are high and large volumes of reagent and by-product have to be transported to and from the plant. Although the gypsum product is saleable, its selling price is low, and is in competition with naturally occuring gypsum, which is easy to mine. The other by-product is crystalline calcium chloride which has few commercial uses and has to be disposed of.

The lime spray-dry process

This process is also in commercial use and is particularly useful for power stations that burn lower sulphur coal (< 1.5% S). The hot flue gas is sprayed with a fine lime slurry in a chamber, producing a dry powder product. Desulphurisation efficiency is low, between 75 and 80 per cent. The lime reagent is produced by burning limestone

($CaCO_3$) in a kiln, then mixing with water to make a slurry. The dry product is removed as a powder from the bottom of the chamber by pneumatic or mechanical handling equipment. The powder contains the sulphur, chloride and other impurities from the flue gas and has no large scale commercial use. The only practical option is to dump it as landfill, and precautions must be taken to prevent leaching of the chemicals in the product into the surrounding ground.

The lime spray-dry process also suffers from a few process problems. The main one is the problem of obtaining sufficient contact time between the lime particles and the gas. There is no recirculation loop, as in the limestone gypsum process, so the gas path and lime spray pattern must be carefully designed to provide a sufficiently long reaction time. This is difficult, which is why high desulphurisation efficiency is hard to achieve. A further problem results from the properties of the dry product removed from the spray-dry chamber. Although easy to handle when dry, the powder is *pozzolanic* – it sets solid if it gets wet – so it must be handled and transported dry. Mechanically, the design of the lime-slurry spraying equipment is important. A rotating atomiser is used, situated in the top of the spray-dry chamber. If this wears or suffers blockage, the spray pattern is affected, which reduces further the desulphurisation efficiency. One advantage of the process is the lack of serious material corrosion problems. There is some erosion, caused by impingement of dry particles on various items of equipment, but these problems are easily resolved by using rubber or plastic linings.

Capital costs of a lime spray-dry FGD plant are low, as the equipment is simple. The running cost, however, is higher, due mainly to the high cost of the burnt lime reagent (three or four times that of quarried limestone) and the landfill disposal cost of the dry powder by-product. The total lifecycle cost for a fixed size of power plant is 30–35 per cent lower than for other FGD processes, but this must be offset against the lower desulphurisation efficiency, so the lifecycle cost per kilogram of sulphur removed from the gas is about the same. There are 20–30 large plants in commercial operation in various countries and most have worked well, unlike some other systems. There is little development of the lime spray-dry process – the technology tends to be rather static. Attempts have been made to find a commercial use for the dry powder product; some has been used as a filler material for low-grade building materials and low-strength surfaces such as playgrounds and car parks but these have not been very successful.

The Wellman–Lord (regenerative) process

This is another well-proven process, although used mainly on the gas treatment of steel-making plants rather than power stations. It can achieve desulphurisation efficiencies of more than 90 per cent with high inlet SO_2 concentrations. It is classed as a *regenerative* process because the absorbant can be regenerated and circulated for reuse in the absorber circuit. The process works by mixing the hot gas with aqueous sodium sulphite (Na_2SO_3) which reacts with the SO_2:

$$SO_2 + Na_2SO_3 + H_2O = 2NaHSO_3$$

$$2Na_2SO_3 + SO_2 = 2Na_2SO_4 + S.$$

The sodium bisulphite formed during the reaction process is soluble. This solution of sodium bisulphite is treated (regenerated) in an evaporator to make water, concentrated SO_2 gas and reconstituted sodium sulphite which is then reused. Not all can be regenerated, however, so it is necessary to take off a continuous bleed stream which forms a dry crystalline by-product. The SO_2 gas which is produced can be made into either sulphuric acid (H_2SO_4) or elemental sulphur, both of which are saleable as an industrial product. For example, for the production of sulphur

$$3SO_2 + 2CH_4 = 2CO_2 + 2H_2O + 2H_2S$$

$$2H_2S + SO_2 = 2H_2O + 3S.$$

The main process problem with the Wellman–Lord system is that it cannot operate well with a high chloride content in the flue gas. A separate prescrubber is needed. The flue gas is cooled by the absorption process so a reheater must be installed between the absorber vessel and the chimney. Materials of construction do not present a major problem – suitable materials are well known from their use in process chemical plants.

The by-product, as mentioned, can be either elemental sulphur or sulphuric acid. Elemental sulphur is a difficult product to deal with; it has to be transported in its molten state, so long-term storage is unrealistic. Once solidified, it can be pelletised and sold. To make sulphuric acid an additional plant is required. Sulphuric acid is widely used as an industrial feedstock but it is an easily available substance so the market price is low. The Wellman–Lord system has a high capital cost (because the plant is complex) *and* a higher energy consumption than the other commercial FGD systems. Although this can be offset partially by the sale of the sulphur by-product, its lifecycle cost is still high.

Comparisons

We can look more closely at some comparisons between the FGD processes. These are broad comparisons and are more useful as a tool to compare the main classes of FGD technologies rather than the many variants that have been developed within each class. The easiest comparison to make, because it is the most visible, is the quantity of reagent and by-product materials that must be transported to and from the FGD plant. This is an important consideration from a logistics/transport management and a planning consent viewpoint – it is also part of the *context* part of the total design model. Figure 18.4 shows the amount of material transport needed for a 2000 MW power station – there are about 20 of these in the UK. The material quantities are shown in comparison to the pre-FGD amounts of coal and ash. They are only shown for the limestone–gypsum system – the lime spray-dry process has a much smaller reagent requirement due, in part, to the smaller amount of sulphur that it is able to remove from the gases. Logistically, the availability of reagents does not present much difficulty; the main problems are caused by the disposal routes for the various by-products. Although the limestone–gypsum and Wellman–Lord processes produce useable by-products, the blunt economic facts are that the products are easily obtainable from alternative sources and so the 'marketability' of the FGD-derived product is not good. The important thing is for the power station to get someone to take the product away rather than make a big profit on it. This is *broadly* the situation in the USA, Germany and particularly Japan, where FGD systems have been installed for over 20 years.

A second comparison point is the state of *technical development* of the main FGD classes. The three main generic FGD process have all been in commercial operation on large power stations but have not *developed* to the same extent. Figure 18.5 summarises the relative development stages, specifically for use in plants burning coal containing more than 1 per cent sulphur. Note the slightly different progress through the various stages of development – the horizontal axis represents technological *progress*, not simply timescale. The nature of the progress varies between processes; the development of the limestone–gypsum process, for instance, has been driven partly by the requirements of the gypsum industry who will only buy a dry gypsum containing low quantities of chloride and heavy metal contaminants. Similarly, the lime spray-dry process has seen development programmes designed to try to improve the efficiency of the desulphurisation reactions and so increase the percentage of SO_2 that can be removed from the gas.

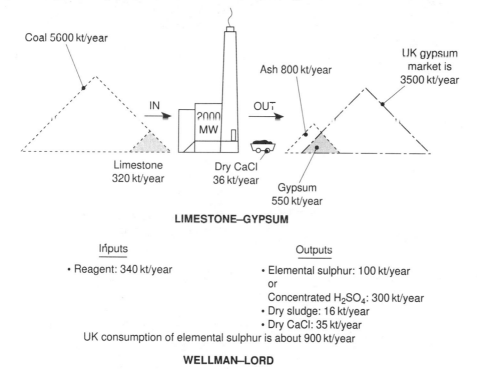

LIMESTONE–GYPSUM

Inputs

• Reagent: 340 kt/year

Outputs

• Elemental sulphur: 100 kt/year
  or
  Concentrated H_2SO_4: 300 kt/year
• Dry sludge: 16 kt/year
• Dry CaCl: 35 kt/year
UK consumption of elemental sulphur is about 900 kt/year

WELLMAN–LORD

Figure 18.4 Material transport and disposal

The third comparison is lifecycle cost, again based on an FGD plant for a 2000 MW power station. For reasons of commercial confidentiality, exact costs of desulphurisation programmes are normally not divulged; it is possible, however, to make an assessment of the relative costs of main processes using data published by electric utilities, the plants' users. Figure 18.5 shows the approximate *relative* lifecycle cost of the main FGD processes. These are based on the assumption of each

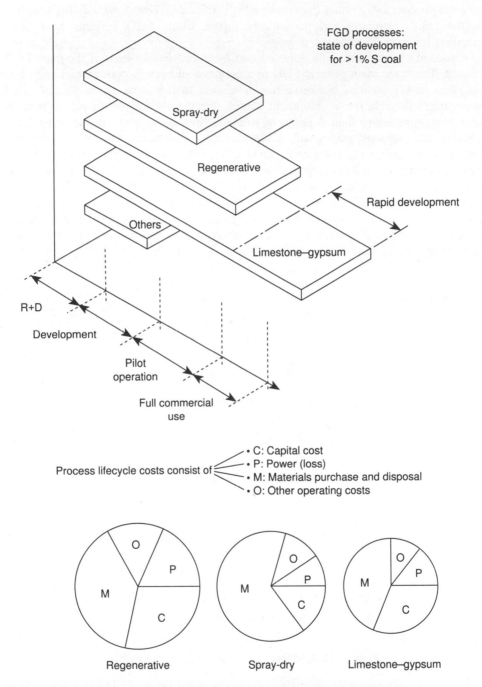

Figure 18.5 Process development and lifecycle cost comparisons

process operating to its maximum desulphurisation efficiency on a 2000 MW power station burning high sulphur coal and so do *not* show accurately the relative costs per unit mass of sulphur removed. The figure also shows roughly the different components of the lifecycle cost, assuming a 15–20 year operating life. As a guideline, the effect of fitting FGD plant to a 2000 MW power station is to increase the cost of electricity generation by about 8 per cent.

18.4 The total design approach

Now we can take a step back and consider the total design approach. I defined this as a way of looking at engineering design in its widest sense, and appreciating the fact that the 'nuts and bolts' mechanics of engineering design do not stand alone as the sole criterion on which design decisions are made. The purpose of the total design approach is to help your understanding of design issues – and to help in making design *decisions and choices*. Practically, design choices, which may be made by a group of people, are helped by documentation such as design reviews, design reports and the like. Ultimately this is what we are working towards by using the total design approach – a mechanism which we can use to help prioritise and clarify design issues. This means reports. In this case study we will show how this approach is used in the choice of FGD plant design. FGD plant, as previously outlined, is based on sound chemical and mechanical engineering principles but its choice of application is affected by important technological and context issues.

A first step

Concepts are fine, but where do we stand? The first step is to look at the representation of the total design process shown in Fig. 18.6. We described total design as:

$$Total\ design = Technology + Engineering + Context$$

It is possible to break each term down, mainly for convenience, into their basic elements relating to FGD plant. Now move on – we need to sort the three terms into order. Figure 18.6 shows them in a 'nested arrangement' with 'engineering' at the centre. The engineering constraints and problems found with any technology have the greatest *inflexibility*. This is because they are the result of physical and chemical laws. You cannot change these, however well versed your knowledge of the technology or the strength of your group on the context. They are immutable – this is why Fig. 18.6 shows the engineering constraints at the centre of the problem. Returning to FGD, we can see that one of the first steps needs to be to find, and then report, these engineering constraints. If you think carefully about the FGD processes, you should see that it is the nature of the *desulphurisation reactions* that produce these engineering constraints. The reactions produce the chemical conditions that promote the process problems (scaling, etc.) and cause the corrosive and erosive conditions that cause the materials problems. You must express the inherent *limitations* of the reactions clearly, or they will be misunderstood.

The middle 'box' in Fig. 18.6 represents *technology* problems. They are represented in this way because they are much 'wider' than the central engineering

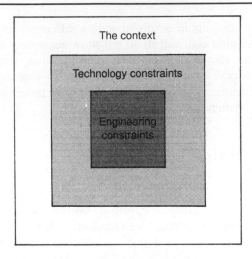

Figure 18.6 The total design model – outline

problems. At first glance it appears that the different FGD technologies are different ways to do the same thing, i.e. desulphurise the flue gas. This is not strictly true – rarely are technologies absolutely equivalent. They rely on sets of physical and chemical laws which do not *allow* variety; you can expect different processes to exhibit fundamental differences – these need not necessarily be obvious but they will be there, if you look. A good example is the way that HCl is treated by the limestone–gypsum FGD system. From the technical outlines given, two points stand out:

- HCl causes problems with desulphurisation reactions at absorber pH values.
- To remove HCl at its optimum pH, a separate prescrubber vessel is needed, which is expensive.

A little lateral (and technical) thought should now raise the question as to whether a prescrubber is essential for the other FGD processes or whether it is really only an option. The answer *may* not be simple, but properly pursued will help add value to your design assessment. Better assessments end in a clearer view of the choices – less is left to chance. This is only an example, but can you see how it has 'opened up' the problem? This is like the concept of 'closed problems' introduced in Chapter 2, although in this case we are looking for additional information that will help us choose between alternatives, rather than look for technical solutions to a specific engineering problem. The general principle of the total design approach is, however, the same.

The outside 'box' in Fig. 18.6 represents the *context* part of the problem. The context contains the widest range of problem issues. Some have a link, of sorts, with the engineering and technology-based problems and can be linked directly to a technical aspect of FGD. Others are less tangible – you could be forgiven for thinking that they seem almost to have a life of their own. This is not unusual. Problems within the context are best found by 'thinking around' the subject, looking for the external issues that will affect the choice of FGD process design. We identified a few of the simpler ones: legislative requirements, planning consents, logistics and commercial issues in Fig. 18.2. We can pull these issues into better focus by posing questions. For instance:

- Supply of reagent materials:
  - Where from?
  - How to transport them?
  - Quality control?

- Disposal of by-products:
  - Quantities?
  - Landfill options?
  - Market options?
  - Environmental problems: leaching, air (dust) pollution, long-term stability?

These are only examples, but even at this low level of analysis, we have already started to deal with the supply, transport and disposal issues which are of key importance for local authority planning consent.

There is one important point about looking at the context part of a total design approach: it is wise to treat the context as a set of *constraints*; don't be misled into thinking that it can act as a 'driving force' towards the choice of FGD process for a specific power station application. The main driving forces come from the centre of Fig. 18.6; the outside 'context box' is best thought of as containing obstacles (*flexible obstacles*, remember) that have to be overcome. Try to keep everything in focus.

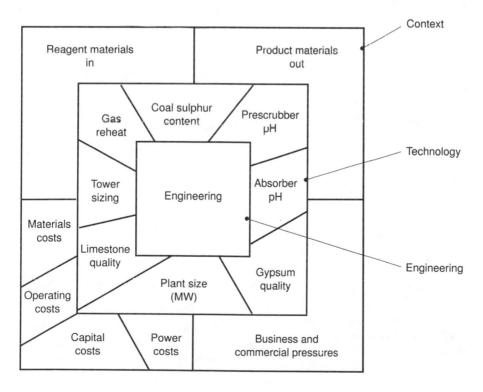

Figure 18.7 The completed FGD total design model

Finally, we need to look at the thorny issue of business and commercial pressures. There is a simple way to treat these:

- It is essential to focus on *costs*; but
- don't get too involved in trying to anticipate the business and commercial pressures that you think lie behind them.

Cost estimates are important and should *always* be included as part of a total design approach. Don't worry too much about the accuracy of your estimates (everyone else's are inaccurate as well) but do make sure that you have used a lifecycle cost approach – capital costs alone are only part of the story. As for the business and commercial pressures that lie in the background, they are not a useful part of the total design approach. Most of them are hidden, and they change unpredictably, and often on a daily basis. Leave them to others to deal with and don't let them cloud the

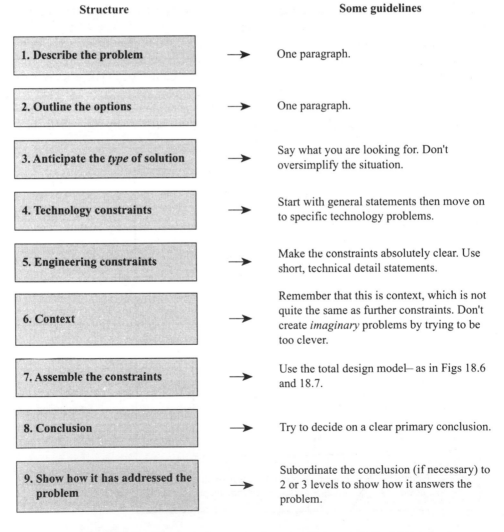

Structure	Some guidelines
1. Describe the problem	One paragraph.
2. Outline the options	One paragraph.
3. Anticipate the *type* of solution	Say what you are looking for. Don't oversimplify the situation.
4. Technology constraints	Start with general statements then move on to specific technology problems.
5. Engineering constraints	Make the constraints absolutely clear. Use short, technical detail statements.
6. Context	Remember that this is context, which is not quite the same as further constraints. Don't create *imaginary* problems by trying to be too clever.
7. Assemble the constraints	Use the total design model– as in Figs 18.6 and 18.7.
8. Conclusion	Try to decide on a clear primary conclusion.
9. Show how it has addressed the problem	Subordinate the conclusion (if necessary) to 2 or 3 levels to show how it answers the problem.

Figure 18.8 Total design approach – use this reporting structure

clearer issues that we have teased out of the total design model. As a summary, Fig. 18.7 shows the location of the FGD issues we have discussed, within the three levels of the completed total design model. Such diagrams are often a useful way to summarise complex design situations and to help focus on the issues that influence the choice between alternative designs.

The purpose of all this background is to help you work towards a reporting structure, the purpose of the reports being to help, and promote, the choice of which FGD process to use for a specific application. The readership will predominantly be among the plant users but such reports are also read by financing banks, local authority planning departments, water authorities and other less directly connected parties who have an interest in the type of FGD plant chosen and how the choice was made. Which is the best way to approach these reports – should they be long, decisive or woolly, technically comprehensive or superficial? Figure 18.8 shows a basic structure which works well for broad 'total design' types of problems. You can use this as a system of reference points to help guide the context of the report and keep it in a logical order. Note how the sections broadly follow the order of 'nesting', i.e. technology, engineering, context. This makes for a better report.

18.5 Case study task

This case study task is designed as a reporting exercise. The objective is to apply the FGD scenarios discussed in the text to the choice of FGD process design for a new 2000 MW coal-fired power station which is designed to burn 1.5% sulphur coal. The task is best addressed as a group exercise.

- First, read the case study text to get an overview of the FGD technologies and the principles of the total design approach.
- Discuss in groups the main issues of *design choice*. The group approach, whether arranged formally or informally, is the best way to gain the wide overall view required.
- Draft a short report explaining the main issues and drawing conclusions on the choice of FGD process. The report should follow the structure shown in Fig. 18.8 and be no longer than 750 words. Simple diagrams can be included if you feel they help explain things in a clear way. The report content needs to be carefully chosen to suit the general readership mentioned in the case study text and include *clear* conclusions and recommendations. If it is necessary to divide responsibilities for sections of the report within the group, try to use the three elements of the total design approach discussed in the text: Engineering, Technology and Context.

19 The fast yacht *Dying Swan* – complex failure

Keywords

Novel design – 'functional' specifications – cooling system design – drive train alignment – lubricating oil flows – thrust bearings – dynamic effects – thermal expansion – poor design interfaces and synthesis – human error – complex failure – diverse information – project management.

19.1 Objectives

Unfortunately, not all engineering failures are straightforward and simple to understand. The problem of *complexity* introduced in Chapter 2 manifests itself in many failure cases. The root of this complexity lies with the nature of engineering designs themselves – even the simplest machine comprises a complicated mix of engineering disciplines and interlinked 'systems'. There are, therefore, many opportunities for things to fail, often as the result of a combination of events, rather than a single identifiable cause. The main objective of this chapter is to show what a *complex failure* design problem looks like, and to demonstrate some of the difficulties of understanding what can be a rather involved and often confusing picture. The chapter also demonstrates the type of methodology that can help you *solve* complex failure and redesign problems.

You can expect this case study to be complicated and multidisciplinary. Fortunately, the relevant engineering principles and calculations themselves are not particularly complicated; it is their *interlinking* that can be confusing. The case is about a new design of fast, gas-turbine powered yacht: the *Dying Swan*.

19.2 The *Dying Swan*

The 'fast' yacht *Dying Swan* can perhaps be considered as something slightly more than just a yacht. At 84 metres in length, with purposefully flared bows, raked stern and a top speed in excess of 50 knots, it is effectively a small destroyer in civilian disguise. It is smooth, white and undeniably aesthetic – and one of the fastest vessels of its type ever built. Unfortunately, it was designed in rather a hurry – a number of the equipment contracts had been let in a rushed fashion and a relaxed view taken of several of the design interfaces in the machinery systems. It would be safe to say that the issue of *design synthesis* was not exactly the most popular subject on the lips of the naval architects and brokers as they pored over their glossy design submissions. Figure 19.1 shows a similar vessel, of about the same size, but from a different manufacturer.

Figure 19.1 A typical 'fast yacht'

For a large non-planing vessel to reach speeds in excess of 50 knots the power-to-displacement (weight) ratio must be extremely high. Because of the correspondingly high fuel consumption at this speed the *Dying Swan* has a composite propulsion plant comprising two high-speed diesel engines (5 MW each) and a single high-speed, 7000 rpm, 15 MW gas turbine (GT) used as 'boost power' to achieve the higher speeds. The diesels and gas turbine drive waterjet propulsion units, instead of conventional propellers. The GT, because of its high (reject) heat output, has its own discrete cooling system, separate from the others in the vessel.

The GT unit spends most of its operating hours at full power. The 'run-up' from cold is fully automatic and 100 per cent power can be achieved in seven minutes. Shut-down is similarly controlled. Because of the high GT fuel consumption, periods of full 'boost' power (50 knots) rarely last for more than three hours. During these periods the GT cooling system temperatures do not always achieve steady-state conditions, particularly when the seawater temperature is higher than 20°C. The propulsion plant is specified to operate with a maximum seawater temperature of 30°C.

19.3 The problem

Following an extended period of low-speed cruising using its diesel engines, *Dying Swan*'s gas turbine was started up, using automatic control, for high-speed running. The sea temperature was 23 °C. At 22.00 hours the GT reached full power, a documented 15 MW, and the yacht's engineers retired to the soundproofed control room for protection against the high noise levels. At 22.10 hours, the gas turbine lubricating oil (GTLO) high-temperature alarm sounded. An engineer responded to the alarm and entered the machinery spaces to make some 'manual adjustments'. After 45 seconds, it was assumed that the GTLO temperature had reduced to below the set alarm level because the warning light in the control room was seen to go out.

At 22.17 hours, without any further warning, or alarms, the GT gearbox started to produce violent 'hammering' noises, accompanied by increasing vibration. In the 15 seconds before the GT trip operated, the gearbox casing region nearest the GT was seen to glow red, as the noise increased. When the unit eventually ground to a stop it remained too hot to touch for two hours. The *Dying Swan* returned to port, at 12 knots, using its diesel engines.

Upon removal of the gearbox casing the gears themselves were seen to be apparently undamaged. The input pinion thrust bearing however was a different story – it was totally destroyed – all that was left of the thrust pads, cage and collar was burnt and twisted metal.

The follow-up was not quite so devastating in its clarity. Almost everyone blamed the gearbox manufacturers, who reported that they had built several gearboxes like this one and 'the others were all right'. They did however mention the incident to the gearbox designer, who felt that it must have been filled with the 'wrong oil'. The oil manufacturer, naturally, did not agree. A lateral-thinking yacht's engineer surprised everyone (initially) by explaining that it was not actually the *gearbox* that had failed, it was the input pinion thrust bearing. In turn, the thrust bearing manufacturer, on being told that the gearbox manufacturers had been told by the oil manufacturer that his oil 'was not to blame', concluded vehemently that it must be something to do with excess axial thrust from the single helical gears or the flexible coupling. Increasingly elaborate circular arguments went on late into the night . . . .

Problem: insight

There has clearly been a major machinery failure which occurred whilst the vessel was operating within its specified design parameters of speed, power and sea temperature. The yacht is new, so a wear and tear problem or any type of 'progressive deterioration' cause is unlikely. It is interesting that the reliability of steady state (full speed) performance of the gearbox and associated lubricating oil systems is actually unproven. The vessel is also of novel and advanced design, so design interface problems cannot be ruled out. Initial responses and 'snap judgements' from the various machinery designers and manufacturers are not untypical – perhaps what you would expect under the circumstances. What is clear is that they do little to help solve the problem, so it makes sense to consider them of peripheral interest at this early stage.

It is unlikely that simple replacement of the failed thrust bearing with an identical new one will be a satisfactory solution. Until you have identified the root causes of the failure (and have firmly eliminated other possible contributory factors) you have absolutely no guarantee that a new bearing would not fail in the same way as the original one.

19.4 Technical design

The layout of the vessel and its GT propulsion plant is shown in Fig. 19.2. The GT drives the centre waterjet via a double reduction single helical gearbox, which reduces rotational speed from 7000 rpm to 430 rpm. The gearbox and its associated lubricating oil (LO) system pipework occupy a space approximately 3 m × 2 m × 2 m within the tight confines of the vessel's machinery space.

The LO system layout

The LO system is arranged in a parallel configuration as shown in Fig. 19.3. The gas turbine gearbox lubricating oil (GLO) is drawn from a sump tank through suction filters and serves the dual purpose of lubricating the gearbox *and* providing a cooling source for the gas turbine lubricating oil (GTLO), which is contained in a separate closed circuit. A manual diverter valve is installed to control distribution of flow between the two 'arms' of the circuit.

Oil pressure is provided by two positive displacement gear pumps of similar ratings arranged in parallel. During full speed operation, the main gearbox-driven pump provides the necessary system pressure. The electrically driven auxiliary pump cuts in automatically during start-up and shut-down conditions when the gear-driven pump is running below rated speed and pressure falls below 2 bar (gauge). Both pumps draw from a common suction pipe extending from the bottom of the atmospheric sump tank located underneath the gearbox. At 'full oil flow' this sump tank has a residence time of approximately ten minutes to allow the LO to de-aerate.

Temperatures and pressure are monitored at several points in the LO system. Simple mercury thermometers and Bourdon-type pressure gauges are used in the various cooler inlet/outlet connections whilst internal gearbox temperatures are sensed using bimetallic thermocouples, connected to the vessel's monitoring and alarm system. There is no 'as installed' method of measuring oil flow 'volume' (or mass) in any part of the system – the only way to determine the volume of oil flowing at any operating condition is by assessing the resulting effects on the pressures and temperatures.

Gearbox LO distribution

Oil enters the gearbox through a 90 mm i.d. manifold which then subdivides into several systems of smaller diameter pipe. Notwithstanding the actual pipe diameter used, internal LO distribution within the gearbox is governed by fixed orifices installed as shown in Figs 19.4 and 19.5. These orifices are factory-installed by the

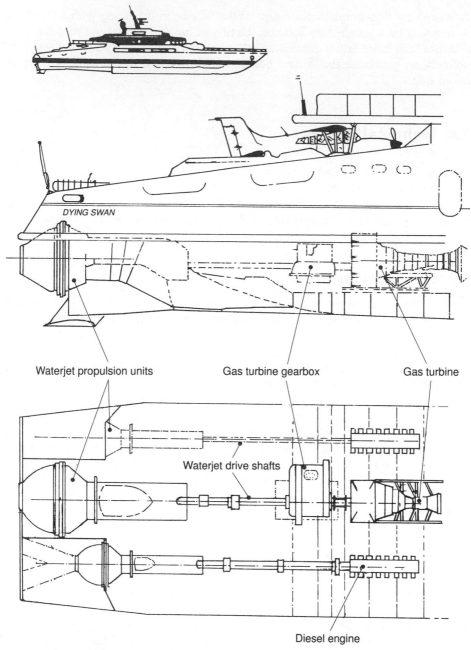

DYING SWAN

Waterjet propulsion units   Gas turbine gearbox   Gas turbine

Waterjet drive shafts

Diesel engine

Figure 19.2 Fast yacht *Dying Swan* – plant layout

gearbox manufacturer, based on the various bearing temperatures experienced during the factory running test, which is carried out in the 'no-load' condition.

Manufacturer's data for the gearbox type indicates an approximate internal oil distribution as shown in Fig. 19.5 but emphasises that these values should be treated as 'indicative only' and not necessarily representative of an individual gearbox.

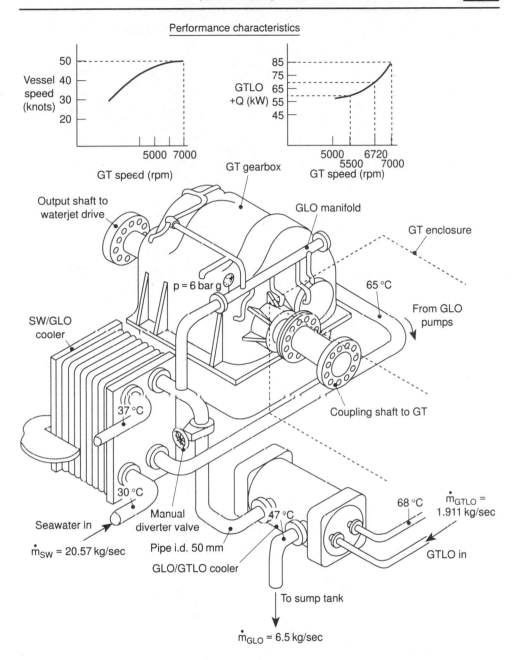

Figure 19.3 GT gearbox cooling circuit arrangement

Based on these indicative figures, the overall efficiency of the gearbox is stated as being 97.38%. Gearbox LO inlet conditions are clearly specified as 4.0 bar (gauge) minimum pressure and a maximum temperature of 48°C.

As the gearbox-driven LO pump is of the positive displacement type, discharge pressure can be set at whatever pressure is required by adjusting the relief valve setting. The mass throughput of the pump is broadly proportional to its rotational

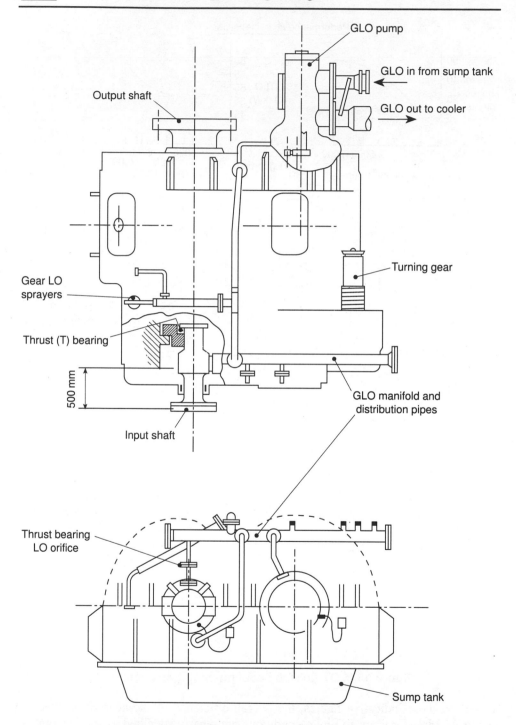

Figure 19.4 GT gearbox general arrangement

speed as shown in Fig. 19.6. The pump is self-priming and will pump successfully small amounts of any air that become entrained in the oil.

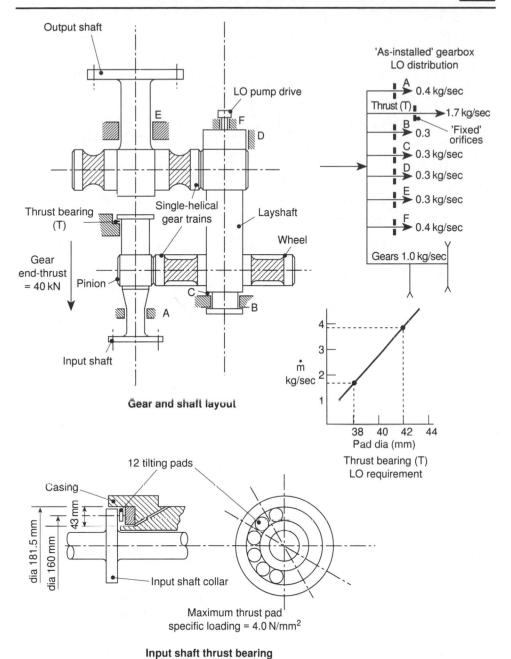

Figure 19.5 Gearbox LO distribution – thrust bearing

Cooling capability

The only heat sink for the system is provided by the seawater/gear oil (SW/GLO) cooler, which is a single-pass parallel flow design as shown in Fig. 19.6. This cooler has to dissipate the combined heat load transferred by the GLO, GTLO and LO pump

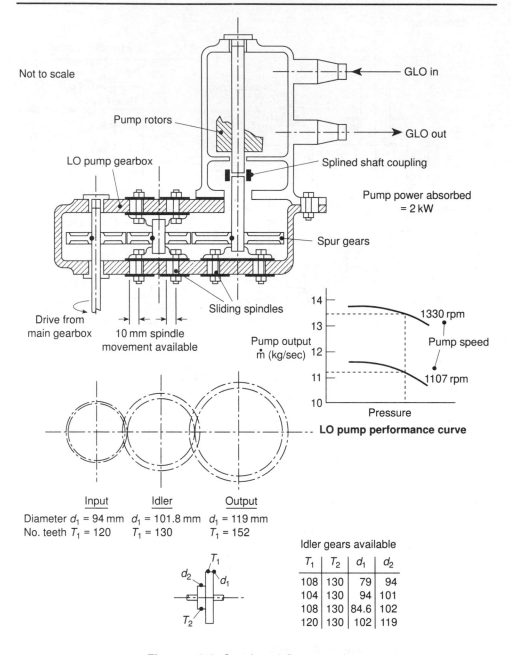

Figure 19.6 Gearbox LO pump data

under constant full speed conditions in the highest specified sea temperature. Being seawater cooled, the cooler is subject to fouling; however the manufacturer's data sheet for this particular design clearly indicates a nominal 'cooling capacity' of 600 kW. A thermostatically controlled bypass valve restricts seawater flow during conditions of low seawater temperature and/or low GT speed to maintain the GLO above the minimum temperature necessary to avoid condensation and thermal shocks in the gearbox.

The GLO/GTLO cooler is of 'compact' design to fit into a tight physical space. The arrangement comprises a single shell pass and two tube passes. The 'best' operational figures recorded for this cooler are as shown in Fig. 19.3. The GLO inlet pipe to the cooler has an external diameter of 50 mm, which looks 'rather small' compared to the overall dimensions of the cooler. Manufacturer's data for the cooler performance is based strongly on empirical results because of the difficulties in analysing accurately the heat transfer characteristics of this type of design (see Fig. 19.7).

GT drive train alignment

The output shaft is connected to the input pinion shaft of the gearbox by a 'high speed' coupling shaft. This has flexible diaphragm elements and so allows some lateral misalignment between the GT and gearbox but, more importantly, allows axial tension ('prestretch') to be incorporated when the unit is in the cold (ambient temperature) stationary condition. This axial prestretch is necessary to compensate for the thermal growth of the coupling shaft and the gearbox input pinion shaft when the unit reaches full speed operating temperature.

Figure 19.8 shows the coupling shaft arrangement. The axial thrust bearing, located on the gearbox input pinion shaft, carries the combined axial force (in a forward direction towards the GT) from the single helical gear train end thrust and that resulting from the cold prestretch of the high-speed coupling shaft. In the hot condition, at full operating power (steady state condition) an axial *tension* must be maintained in the drive train. This is because the output shaft bearing is designed to run with a tension loading – if it is unloaded, or a compressive (aft) load is applied, the bearing will have a very short life.

During run-up of the GT the drive train only spends a very short time in the cold, full speed (7000 rpm) condition. The shafts quickly heat up to 120°C. Because of centrifugal effects on the flexible diaphragm elements in the high-speed coupling shaft the tension force in the coupling varies with rotational speed. The relevant characteristics, based on 'type-test' data for similar couplings, are shown in Fig. 19.8. The axial prestretch of the coupling can be adjusted by moving the GT forward or aft. The mounting system comprises a series of high-tensile linkages arranged as a three-dimensional spaceframe. During any axial adjustment of the GT position its lateral alignment, relative to the forward/aft axis of the vessel, must be maintained, otherwise the output bearing and gearbox input bearing may fail due to excessive lateral forces and vibration.

Thrust bearing design

Figure 19.5 showed the outline design of the 'Michell type' tilting pad thrust bearing fitted to the gearbox pinion input shaft. This is the bearing that failed. Twelve circular pads of 40 mm diameter, each with a stepped backing (centre-pivot), are located in a rotating cage. Oil is fed from the gearbox LO distribution manifold through a fixed 11 mm orifice and then into the bearing housing channels which pass oil on to the bearing surfaces. It then drains to the sump. The load bearing capability

SW/GLO cooler – parallel flow

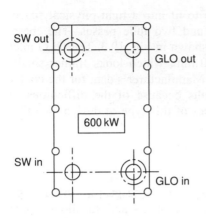

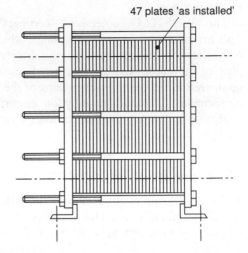

47 plates 'as installed'

SW out

GLO out

600 kW

SW in

GLO in

$Cp_{SW} \cong 4200$ J/kg K
$Cp_{GLO} \cong Cp_{GTLO} \cong 2350$ J/kg K
$\rho_{GLO} \cong \rho_{GTLO} \cong 840$ kg/m³
$U \cong 250$ W/m² K

GLO/GTLO cooler – 'compact design'

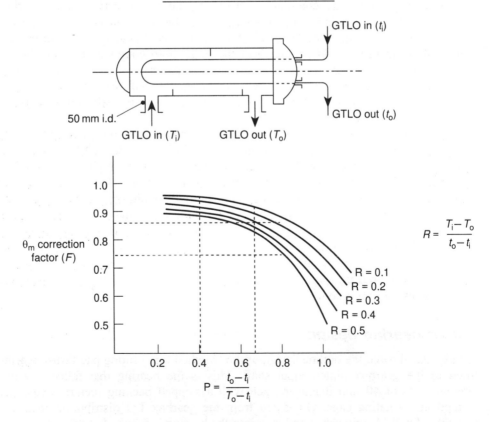

GTLO in (t_i)

GTLO out (t_o)

50 mm i.d.

GTLO in (T_i)

GTLO out (T_o)

θ_m correction factor (F)

R = 0.1
R = 0.2
R = 0.3
R = 0.4
R = 0.5

$R = \dfrac{T_i - T_o}{t_o - t_i}$

$P = \dfrac{t_o - t_i}{T_o - t_i}$

Figure 19.7 Cooler data

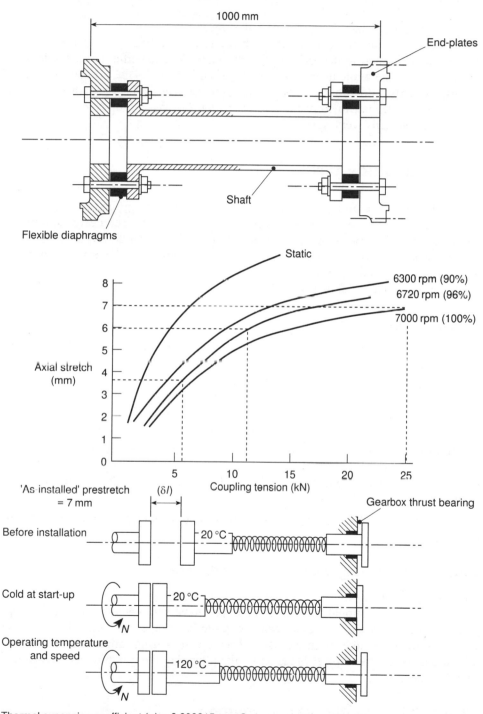

Coupling shaft assembly

1000 mm

End-plates

Flexible diaphragms

Shaft

Static

6300 rpm (90%)
6720 rpm (96%)
7000 rpm (100%)

Axial stretch
(mm)

Coupling tension (kN)

'As installed' prestretch
= 7 mm

(δl)

Gearbox thrust bearing

Before installation

20 °C

Cold at start-up

20 °C

N

Operating temperature
and speed

120 °C

N

Thermal expansion coefficient (α) = 0.000015 per °C

Figure 19.8 Coupling shaft arrangement and characteristics

of the thrust bearing is determined by the maximum pad specific loading, which is in turn influenced by the LO film thickness which forms in use. Manufacturer's data (largely empirically based) is shown in the figure.

Temperature monitoring of the thrust bearing, as installed, is limited to a single thermocouple located in the drain pipe to the sump. The maximum acceptable metal temperature for the white metal bearing alloy used is 130°C – the primary purpose of having a sufficient flow to the thrust bearing is to remove the heat generated from the pads, the flow needed to form the hydrodynamic oil wedge between the pads and collar being actually very small.

19.5 Case study tasks

This is designed as a group exercise with one member of the group being designated as Project Manager. Your collective task is to specify and report a design solution to the problems experienced on board the yacht *Dying Swan*. To do this successfully it is first necessary to investigate the design parameters and interfaces using the information given in the case.

The task consists of four essential phases.

Phase 1: the design check

Check the design parameters of the 'as installed' systems, starting with the operational data given in the case and then testing against accepted technical design formulae and rules. On the basis of the results, decide whether the installed equipment and systems are adequate for their required task. Note that some technical parameters are interlinked with those from other 'systems'. For example:

- The thrust bearing pad size affects the required thrust bearing LO flow.
- The thrust bearing LO flow has an influence on the distribution in the other parts of the gearbox.
- Higher thrust pad loadings generate more heat which has to be dissipated by the 'heat sink' SW/GLO cooler.

A proforma (Fig. 19.9) is provided to help you organise data collection and calculations.

Phase 2: resolving technical uncertainties

Any outstanding technical uncertainties about the 'as installed' design of the machinery systems should be decided after you have completed the phase 1 design check. It is not *necessarily* the case that the design parameters of some of the systems that seem to be 'remote' from the issue of the thrust bearing failure are satisfactory. Try to understand the Project Manager's role in helping the group to take a balanced view of these areas.

Phase 3: specify the design solution

Once you have gained an understanding of the technical characteristics of the case, your collective task is to specify the design solution. Figure 19.9 acts as a proforma

DESIGN CHECK

Overall LO system

LO flow m =	. . . kg/sec
Heat Sink $(-Q)$ =	. . . kW
LO Pump $(+Q)$ =	. . . kW
Gearbox $(+Q)$ =	. . . kW
Overall Q margin =	. . . kW
Gearbox (min) LO m =	. . . kg/sec

Drive train

Cold prestretch =	. . . mm
Max tension (cold) =	. . . kN
Thermal growth =	. . . mm
Max tension (hot) =	. . . kN
Thrust brg area =	. . . mm^2

Notes: LMTD $\theta_m = \dfrac{\theta_1 - \theta_2}{\ln(\theta_1/\theta_2)} \times F$ (factor)

SW/GLO cooler

Q =	. . . kW
θ_m =	. . . °C
A =	. . . m^2
U =	. . . W/m^{2}K
Min pipe dia =	. . . mm

GLO/GTLO cooler

Q =	. . . kW
θ_m =	. . . °C
A =	. . . m^2
U =	. . . W/m^{2}K
Min pipe dia =	. . . mm

Gearbox LO distribution

Bearing	m kg/sec	%
T		
A		
B		
C		
D		
E		
F		
TOTAL		100%m

DESIGN SOLUTION : OUTLINE DATASHEET

- GT speed — rpm
- GTLO $(+Q)$ — kW
- Gearbox ζ — %
- Gearbox $(+Q)$ — kW
- LO Pump $(+Q)$ — kW
- Overall Q margin — kW
- Coupling prestretch — mm
- Max axial tension — kN
- LO circuit arrangement — series/parallel
- Gearbox LO m — kg/sec
- Thrust bearing orifice — mm dia

- SW/GLO cooler
  - Area A — m^2
  - LMTD (θ_m) — °C
  - Configuration — parallel/counterflow/mixed
- GLO/GTLO cooler
  - Area A — m^2
  - LMTD (θ_m) — °C
  - Configuration
- Thrust bearing LO flow m — kg/sec
- Thrust bearing loading — N/mm^2
- Thrust pad dia — mm

Figure 19.9 Datasheet – design check and design solution

that you can use as guidance. You can expect some simple iteration to be involved, because of the interrelated problem areas, but the extent of this iteration depends almost entirely on the way that you approach the problem. Try to follow the outline methodology in Fig. 19.10 as closely as you can.

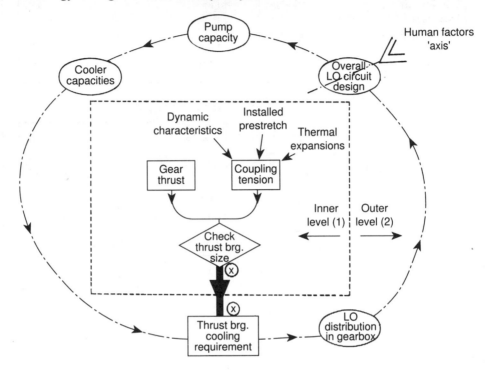

Figure 19.10 Case study methodology

Phase 4: reporting

The group Project Manager's task is to produce a report explaining, in simple terms, what caused the failure on board the *Dying Swan*. Conclusions must be explained in a succinct way and related to actual information provided in the case text and figures. Equally, the various elements of the *design solution* must be presented and explained in a way that shows they are an answer to the failure problem that occurred. It is of little use specifying a design solution, however elaborate, that does not address accurately the fact that a failure of a specific component *has* occurred. Your report must be less than 450 words and be text only. No diagrams. It will be read by the yacht's owner, who is not an engineer, but will also find its way to the shipbuilder and equipment manufacturers – so it must be technically consistent. Try to incorporate at least the following sub-headings, but not necessarily in this order:

- Proposed design solution
- The failure event (description)
- Design review
- Cause of failure.

The project manager's brief

As Project Manager you are fortunate in being party to several quite high-level discussions with managers from various equipment manufacturers. You have picked up some useful technical pointers at these meetings.

Gearbox efficiency It took you some probing to obtain tacit acceptance from the gearbox manufacturers that their quoted efficiency failure of 97%+ is, at best, optimistic for full-load conditions. You agreed that 96.3 per cent was perhaps a more accurate figure to use, even when ignoring the effects of windage losses.

LO pump The gear-driven LO pump installed can sometimes draw in air through a worn suction seal, effectively mixing about 4 per cent of air with the oil throughput. However, you don't know whether or not this happened in this specific case.

Thrust bearing maximum loadings The manufacturer *reluctantly* concedes that his published data on maximum specific pad loadings strictly only applies to thrust pads with offset pivots and that the centre-pivot type should have an additional margin of 20 per cent built in for 'good measure'. He maintains however that the original bearing was still big enough for its purpose.

The owner One of the more unfortunate aspects of your role as Project Manager is your direct liaison with the yacht's owner. His view is that he has just paid over twenty million pounds for a prestige yacht and it is not working.

> 'So why won't it run properly at 50 knots?' The owner continued, 'Look, if it was the crew's fault then just make sure that they can't do it again. Oh, and I'm not paying dry-dock fees so you can't fit any new seawater pumps; you'll have to make do with whatever seawater is flowing around in there already'. He concluded, 'I'm going to have to pay for these repairs and so just make sure that you keep the cost down – stick to a ten per cent design margin or, er, whatever it is that you people normally use'.

The line went dead as you jotted down the key points.

As Project Manager it is your role to work with the other members of your 'design team'. It is not necessary for you to become involved in the various detailed technical calculations but it is important that you obtain an awareness of the results, in order to understand the impact they have on the overall design solution. You are responsible for managing the implementation of the suggested methodology. A key part of this will be ensuring that the various analyses are kept at the right level – neither over-complicated nor too simplified. You need to be prepared to make decisions and resolve technical conflicts.

The technical solution to any multidisciplinary design problem must be carefully organised. Do not expect that it will somehow 'manage itself' – it probably won't. Be careful to ensure that the various technical solutions are all based on clear observations, or robust calculations, and that they all fit together without introducing technical contradictions.

Another of your Project Management responsibilities is to *report* your team's agreed design solution. You should be able to defend its content against technical criticism, showing that it addresses properly the proven causes of the failure. You must also take into account the very clear set of practical limitations set out by the yacht's owner.

Methodology

The only purpose of a methodology is to help solve the problem by generating a workable design solution. We discussed in Chapter 2 that it is often necessary to 'open up' a problem before solutions will start to become apparent – the case of the *Dying Swan*, in common with some other (but not all) design investigations, fits into this category. Two specific points from those generated in Chapter 2 (refer back to Fig. 2.5 if necessary) have particular relevance to this design case:

- The importance of background information – this is because of the multi-disciplinary nature and technical 'width' of the problem. You will see how the issues of LO system cooling capacity, drive train forces, thrust bearing load capability and human intervention are interlinked. Rich background information is almost essential in this type of situation.
- The *active* generation of a design solution – using (nearly always) a combination of robust theoretical analysis and *inductive* reasoning. *Inductive* means thinking for yourself. Try to see the issue of overall LO circuit arrangement in this context.

Figure 19.10 shows a suggested methodology for this design case. Its primary purpose is to show a general flow path applicable to both the 'design check' and 'solution' activities – try not to view it too generally, though. You should draw particular technical inferences from each individual annotation. Note how the methodology is assigned two discrete levels, the 'inner resolution level' addressing the issues of coupling shaft prestretch and thermal growth, and the 'outer level', the wider (but in this case slightly simpler) questions of LO circuit design and its cooling capacity. You should recognise this as a close adaptation of the common 'multilevel' approach suggested in Chapter 2, Fig. 2.6. Note the corresponding existence of the pivot link x—x, in this case the determination of thrust bearing load capability, which acts as the contact point between the two levels of analysis.

At level 1, the gear train end thrust and coupling shaft tension are examples of discrete technical areas that have a sound theoretical origin. It is wise to analyse these first, before moving to the wider level 2 issues.

19.6 Nomenclature

A	Area of heat transfer surface
GLO	Gearbox lubricating oil
GT	Gas turbine
GTLO	Gas turbine lubricating oil

LMTD(θ_m)   Log mean temperature difference $= \dfrac{\theta_1 - \theta_2}{\ln(\theta_1/\theta_2)} \times$ Factor (F)

LO   Lubricating oil

$\pm Q$ Heat transferred (joules)

SW   Seawater

U Heat transfer coefficient

Conclusion

One of the advantages of using a case study approach to teaching (and learning) engineering design is the way that the various principles can be presented in manageable pieces. Hopefully, having worked through the case studies, you can see useful learning points in each. It is not absolutely essential that you work through all the case studies in detail, although this is probably the way to get the best 'value' from their content. A selective approach will still give you an idea of some important principles of the design process.

The case studies should have demonstrated to you the true multidisciplinary nature of engineering design – this was one of the key messages of the book, with several examples of how statics, dynamics, thermodynamics and metallurgy exist together in even the simplest designs. If you have worked through the case study tasks you should also have seen how the multidisciplinary nature of design means that there is rarely one simple solution – this is also an important message: do not waste too much time looking for a perfect, unequivocal answer to design problems.

You may have found areas in some of the case studies where the theoretical aspects are not covered in such great depth as, perhaps, in other design textbooks. This does not mean that the theory is unimportant – just that it can be easily found in other references. In general, these case studies concentrate more on other topics such as costing, quality assurance and design management. These are equally important design issues – you *cannot* afford to ignore them. Lastly, there is no doubt that the case studies, as presented, are imperfect – it has been necessary to omit some details for clarity. Hopefully this should not compromise the usefulness of the exercises for learning purposes – it should make them easier.

The final word

'Exercises, so that's all they were, exercises.'

'It's . . .'

'OK, so they cover a variety of aspects; reliability, quality, strength, fatigue, conceptual design, embodiment, design improvements, managing things, innovation and more innovation, to name but a few.'

'It's a structured trek through the issues . . . of engineering design . . . of how it's done.'

'And the failures?'

'So you can learn from others' mistakes.'

'Makes sense – but what about that stuff on methodologies, on dissecting and structuring problems – did it work?'

'It's not a question of it working, it's a matter of victory or defeat; if you don't have a way of handling the complexity that lives in design problems it will eventually get you. Guaranteed.'

'You're wrong on that one, there are no guarantees – but I think I can agree that you need to put a design problem in order, before you can solve it.'

'Success!'

'So what's the final word then?'

'It has to be *innovation*, I guess.'

'Didn't I read something about the consequences of not innovating?'

'Yes; engineering industry is . . .'

'littered with the bones of companies, and people, that . . .'

'did not innovate.'

'and all had . . .'

'absolutely superb excuses.'

'But they still didn't do it?'

'Right!'

'Solidarity, between us.'

(Together) 'Innovation is the bedrock of the design process.'

Appendix

The technical standards mentioned in this book are available from the following sources:

British Standards Institution (BSI)
British Standards House
389 Chiswick High Road
London W4 4AL

Note that BSI is one of the few organisations in the UK permitted to purchase all international and most national standards direct from source, including ISO, IEC, DIN (English translations), ASME and API.

American Society of Mechanical Engineers (ASME) and related ASTM standards are available from Mechanical Engineering Publications Ltd, Northgate Avenue, Bury St. Edmunds, Suffolk, UK.

American Petroleum Institute (API) standards are available from ILI, Index House, Ascot, Berkshire, UK.

Bibliography

Dhillon, B.S. (1996). *Engineering design*, Irwin, London.

French, M.J. (1985). *Conceptual design for engineers*, Springer Verlag, New York.

French, M.J. (1994). *Invention and evolution: design in nature and engineering*, Cambridge University Press, Cambridge.

Johnson, R.C. (1980). *Optimum design of mechanical elements*, Wiley, New York.

Juvinall, R.C. and Marshek, K.M. (1991). *Fundamentals of machine component design*, Wiley, New York.

Mott, R.L. (1992). *Machine elements in mechanical design*, Merrill, London.

Norton, R.L. (1996). *Machine design, an integrated approach*, Prentice Hall, Englewood Cliffs, NJ.

Pahl, G. and Beitz, W. (1988). *Engineering design. A systematic approach*, Design Council, London.

Pugh, S. (1990). *Total design – integrated methods for successful product engineering*, Addison Wesley, Wokingham.

Shigley, J.E. and Mischke, C.R. (1986). *Standard handbook of machine design*, McGraw-Hill, New York.

Siddall, J.N. (1972). *Analytical decision making in engineering design*, Prentice Hall, Englewood Cliffs, NJ.

Urich, K.T. and Eppinger, S.D. (1995). *Product design and development*, McGraw-Hill, New York.

Index